C0-AWX-536

Energy Strategies for Developing Nations

Energy Strategies for Developing Nations

Joy Dunkerley, William Ramsay
Lincoln Gordon, Elizabeth Cecelski

PUBLISHED FOR RESOURCES FOR THE FUTURE
BY THE JOHNS HOPKINS UNIVERSITY PRESS
BALTIMORE AND LONDON

Copyright © 1981 by Resources for the Future, Inc.

All rights reserved
Manufactured in the United States of America

Published for Resources for the Future
By The Johns Hopkins University Press, Baltimore, Maryland 21218

Library of Congress Catalog Card Number 80-8774
ISBN 0-8018-2596-2
ISBN 0-8018-2597-0 (pbk.)

 RESOURCES FOR THE FUTURE, INC.
1755 Massachusetts Avenue, N.W., Washington, D.C. 20036

Board of Directors: M. Gordon Wolman, *Chairman,* Charles E. Bishop, Roberto de O. Campos, Anne P. Carter, Emery N. Castle, William T. Creson, Jerry D. Geist, David S. R. Leighton, Franklin A. Lindsay, George C. McGhee, Vincent E. McKelvey, Richard W. Manderbach, Laurence I. Moss, Mrs. Oscar M. Ruebhausen, Janez Stanovnik, Charles B. Stauffacher, Russell E. Train, Robert M. White, Franklin H. Williams

Honorary Directors: Horace M. Albright, Erwin D. Canham, Edward J. Cleary, Hugh L. Keenleyside, Edward S. Mason, William S. Paley, John W Vanderwilt

President: Emery N. Castle

Secretary-Treasurer: Edward F. Hand

Resources for the Future is a nonprofit organization for research and education in the development, conservation, and use of natural resources and the improvement of the quality of the environment. It was established in 1952 with the cooperation of the Ford Foundation. Grants for research are accepted from government and private sources only if they meet the conditions of a policy established by the Board of Directors of Resources for the Future. The policy states that RFF shall be solely responsible for the conduct of the research and free to make the research results available to the public. Part of the work of Resources for the Future is carried out by its resident staff; part is supported by grants to universities and other nonprofit organizations. Unless otherwise stated, interpretations and conclusions in RFF publications are those of the authors; the organization takes responsibility for the selection of significant subjects for study, the competence of the researchers, and their freedom of inquiry.

This book is a product of RFF's Center for Energy Policy Research, Milton Russell, director. It was edited by Sally A. Skillings and indexed by Florence Robinson. The figures were drawn by Art Services. The book was designed by Elsa Williams.

RFF Editors: Ruth B. Haas, Jo Hinkel, and Sally A. Skillings

Contents

List of Tables

Appendix Tables

List of Figures

Foreword

The energy situation in the developing countries has quietly assumed major importance. Demand will be growing over the coming decades in these countries even though it may be stagnant in the industrialized nations; higher energy prices there will be taking their most serious toll on economic growth and the environment; and while the ultimate prospects for new supplies of both conventional and nonconventional fuels are bright in many developing countries, timing, costs, and actual outcomes remain uncertain.

The oil price increases that threaten to wipe out years of progress in raising living standards are mostly beyond the control of the developing countries. But how they respond to the changed and changing energy realities will affect their prospects for continued economic growth and in some cases social and political stability. Energy outcomes are crucial to human welfare in the developing world and thus must engage the humanitarian concern of energy exporters and of all energy-importing countries as well.

But the industrialized nations are not mere spectators who can observe with compassionate detachment the energy plight of the developing world. To the extent the demand for oil can be dampened in all countries, there will be less pressure on the world oil market and lower energy prices. Conflicts over energy, which can spill over into other areas, will be muted. To the extent the developing countries can sustain their development momentum, markets for industrial-country exports will be expanded and credit repayment will be easier. Energy outcomes can also affect internal stability and the prospects for peace among the developing countries, and, thus exacerbate or dampen tensions that now bedevil the global scene. Finally, the welfare of the peoples of the developing countries is bound up with our own. Success in mitigating the harm wrought by higher energy costs lessens the already massive problems burdening poorer countries and in doing so also makes the industrialized world more secure.

Doing something about the energy problems of the developing world involves, first of all, understanding what the situation really is. As this book demonstrates, no easy generalizations hold, though some seem to underlie widely advocated approaches to developing-world energy problems. Indeed, the currency of facile generalizations about a topic at once important and complex was one factor that motivated us to launch a broad effort to examine energy issues in the developing-country context. This book is the first major publication to result from this research.

Allan G. Pulsipher, then of the Ford Foundation, was instrumental in initiating the research that resulted in this book. Financial support from the Ford Foundation made the work possible and, melded with RFF's internal funds, provided the impetus to an area of study now supported by a number of sources.

Energy Strategies for Developing Nations is an effort to view energy issues whole. It exposes the complexity of energy and economic opportunities and constraints in the developing countries without exploring all of the passages and byways identified. It offers a way of thinking about energy problems and a framework for guiding further research, some of which is already underway here at RFF.

This book brings forward the issues that must be resolved in formulating energy strategies within developing nations. It helps to set priorities for researchers and policy makers. Forswearing nostrums, *Energy Strategies for Developing Nations* offers a reasoned starting place for those who must think seriously—and then act—at this critical energy policy frontier.

January 1981   Milton Russell, Director,
                                        Center for Energy Policy Research

Preface

The increase in petroleum prices that took place during the 1970s gave rise to widespread problems of economic and energy adjustment throughout the world. Perhaps the widest array of these problems is that facing the oil-importing developing countries.

Higher oil prices lead to accelerated inflation rates, lower growth rates, and balance-of-payments crises. In many developing countries they also increase the pressures on forest and other forms of traditional energy supplies. Superimposed on these problems that demand immediate attention is the equally pressing need to adjust to a new long-term future dependent on other forms of energy. These problems of crisis and transition mean that the task of securing higher living standards for expanding populations—a task which already poses major problems of economic and social management for the governments of developing countries—are being made much more difficult. Energy adds yet another troublesome item to an already overcrowded agenda, and to complicate matters further, this new agenda item cuts across existing governmental organization, requiring radical changes in procedures.

Because of the severity of these problems and also because little is understood in a systematic way about the role of energy in developing areas, we decided to undertake the present study of energy strategies for developing nations. We pose and seek to answer several questions. What is the nature of the energy problem in both traditional and modern sectors of the developing countries? Does the energy crisis constitute a formidable barrier to the continuation of economic development? What options and possibilities are open to the developing nations themselves to solve their problems through adjustments in development strategies and tactics? What is the outlook for increasing supplies of domestically produced fuels—both conventional and "new and renewable"—and which choices are best for which type of country? Improving the efficiency with which energy is used is a lively option in the industrialized world: is it also so for developing nations? And finally, as we believe that there are sufficient areas of common interest among both industrial and developing countries to justify and indeed dictate a common approach to many energy problems, what can international cooperation do to help?

This book is the result of a collaborative effort by the four authors, although each author had responsibility for individual chapters. Thus, the overview, and chapters 1 and 4 were the primary responsibility of Lincoln Gordon. Elizabeth Cecelski wrote chapter 2, Joy Dunkerley chapters 3 and 5, and William Ramsay chapters 6, 7, 8, and 9. All four authors cooperated in the writing of chapter 10. While Joy Dunkerley, Lincoln Gordon, and William Ramsay were co-principal investigators on this project, Joy Dunkerley acted as *prima inter pares* in planning and coordinating the project and the resulting book.

We received help from many sources. The Ford Foundation provided a major part of the financial support for the research and for writing the book. Allan G. Pulsipher, our project officer at the Foundation, gave us much appreciated support and encouragement throughout the project. Within Resources for the Future, Andrew Steinfeld of the Center for Energy Policy Research provided research assistance. John E. Jankowski, Jr., also of the Center, made major contributions to chapter 6 as well as providing wide-ranging assistance in other chapters. Hans Landsberg and Milton Russell, past and present directors of the CEPR, provided steady support to the project, as did Charles J. Hitch and Emery N. Castle, past and present presidents of RFF.

Throughout the project we benefited from the comments, suggestions, and expertise of an advisory group drawn from foundations, independent research institutions, U.S. government agencies, and international organizations. Detailed comments on the manuscript were received from the following members (with their affiliations at the time of the project): Robert Copaken of the Department of Energy, Allan G. Pulsipher of the Ford Foundation, David Hughart of the World Bank, Alan B. Jacobs of the Office of Energy of the U.S. Agency for International Development, Lawrence Ervin of the al Dir'iyyah Institute, James Howe of the Overseas Development Council, Charles R. Blitzer of the U.S. International Development Cooperation Agency, and Robert F. Ichord, Jr., of the Bureau for Asia of the U.S. Agency for International Development. In addition, Edward S. Mason, Professor Emeritus of Harvard University, Kirk Smith of the East-West Center, and Pierre R. Crosson of RFF, also provided extensive manuscript review. Other members of the advisory group and representatives of the energy and development communities gave helpful advice: Marcello Alonso and Phactuel Rego of the Organization of American States, Ali Ezzati and B. J. Choe of the World Bank, Herman Franssen of the U.S. Department of Energy, Michael Gukowski of the United Nations Development Programme, Philip Palmedo of Energy/Development International, C. Anthony Pryor of the Rockefeller Foundation, Eric Melby and Pamela Baldwin of the Office of Energy of the U.S. Agency for International Development, Kathleen Rees of the U.S. Department of Energy, and Henry Kelly of the Solar Energy Research Institute. Sally Skillings, managing editor of the RFF publication staff, had the challenging task of melding a multiauthor manuscript into a consistent whole. Few of the excellent secretarial staff of the CEPR managed to escape some involvement with this manuscript. Their cheerfulness and enthusiasm were much appreciated. The complicated task of manuscript coordination was most efficiently performed by Marilyn D. Meding.

January 1981   Joy Dunkerley
                                                               William Ramsay
                                                               Lincoln Gordon
                                                               Elizabeth Cecelski

Energy Strategies for Developing Nations

Overview

The developing countries constitute an increasingly important component of the global energy scene. With three-quarters of the world's population, they account for one-quarter of total energy consumption and four-fifths of energy exports. Most of them are oil importers who share with industrial countries the twin problems of coping in the short term with sharply higher prices and uncertainties of supply and devising for the longer term the most effective combination of energy supply alternatives to imported oil. For many, there is the further problem of dwindling supplies of traditional forest and farm-based fuels, with severe consequences in environmental degradation and reduced agricultural productivity.

Even before the energy price "shocks" of the 1970s, the developing world faced severe challenges in its struggles for economic progress under conditions of population pressure, capital shortage, paucity of technological skills, and deep-rooted institutional and cultural obstacles to modernization and reform. Effective energy management poses an additional challenge. The impact of energy on capital supplies, the balance of payments, and institutional adaptability have now made it a major potential constraint on development prospects. Because of the intermeshing of energy markets, successes in energy substitution and conservation in developing countries would improve the global energy balance. In greater measure than applies to development in general, the industrial nations share an affirmative interest in mitigating the short-term energy crisis now confronting the developing countries and in securing their full participation in the longer term transition in energy regimes on which the world is now embarked.

1

Energy Supply and Use in Developing Countries

The term "developing countries" (or the South) is applied to some 127 nations, comprising Asia (except Japan), Africa (with South Africa considered a borderline case), and Latin America and the Caribbean. The contrasting group of 27 "industrial countries" (the North) consists of the United States, Canada, Japan, Australia, New Zealand, and Western and Eastern Europe, including the Soviet Union.[1] As of this writing, real incomes per capita in the South average less than one-sixth those of the North.

The developing-country group is much more heterogeneous than the North, with an enormous variety of resource endowments and development conditions. At one end of the spectrum are countries with little economic activity beyond subsistence agriculture and live-stock tending. At the other end are countries such as Korea, Singapore, Brazil, and Mexico with such large and dynamic manufacturing and commercial sectors that they are close to "graduating" into the industrialized category. In between are relatively poor but populous countries such as China, India, and Pakistan where a substantial modern industrial sector is embedded in a primarily agricultural economy. According to most indexes of social development (including those of life expectancy and adult literacy), as well as per capita income, Latin America as a region is well in the lead, followed in turn by Asia and Africa.

These basic conditions are reflected in the patterns of energy supply and use. Figure 0-1 shows the per capita energy consumption levels for each of the developing-country regions and for the entire group, contrasting them with the industrial-country and global aggregates. The narrow dotted bars indicate the population levels for each grouping. The wider bars distinguish between energy consumption from commercial sources (oil, coal, gas, hydroelectricity, and nuclear power) and traditional sources (mainly firewood and charcoal, animal dung, and agricultural wastes). Animate energy is not included. In addition to the sharp contrast in overall energy use per capita between developing and industrial countries—a disparity even wider than that

[1] The summary figures in this overview are based on the United Nations classification that places South Africa and Israel among the industrial countries. The developing-country group comprises the "developing market economies" and the "Asian centrally planned economies," to which data for Taiwan have been added. For further discussion of the definition of "developing countries," see the appendix to chapter 1.

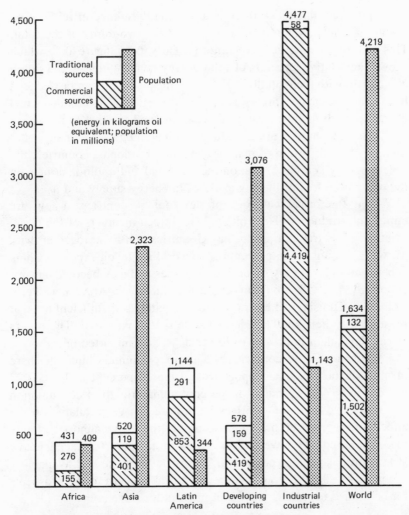

Figure 0-1. Per Capita Energy Consumption and Population, by Region, 1978. Traditional energy data for 1973 from Jyoti K. Parikh, "Energy and Development," World Bank Public Utilities Report No. PUN 43 (Washington, D.C., August 1978), recalculated on the assumption that per capita traditional energy consumption remained unchanged between 1973 and 1978. Commercial energy data for 1978, except for Taiwan, from United Nations, *World Energy Supplies, 1973–1978*, Series J, No. 22 (New York, 1979), with data on hydroelectricity and nuclear energy recalculated on basis of thermal generation primary energy equivalents. Data for Taiwan from an oral communication from the Coordination Council for North American Affairs to Lincoln Gordon. Population data from Population Reference Bureau, *1978 World Population Data Sheet* (Washington, D.C., 1979).

in average real incomes—there is a striking difference in reliance on traditional fuels, which in Africa account for two-thirds of the total. This difference emerges even more clearly from figure 0-2, which shows the relative amounts of total energy consumption in the South and North, with their subdivision into main sources. It can be deduced from the two smaller circles at the top that coal is the dominant fuel in the centrally planned countries of China and North Korea, whose resource endowments are quite exceptional among developing countries. Within the group of market-economy developing countries, the salient energy sources are oil on the one hand and traditional fuels on the other—the foci of their twin crises in energy supply and use.

Within the broad category of developing countries, a few are important producers of petroleum. The fifteen members of the Organization of Petroleum Exporting Countries (OPEC), together with Mexico, account for 86 percent of the world's total oil exports.[2] China is believed to have very extensive resources and may become a major exporter. A small group of countries, including Argentina, Peru, Colombia, Tunisia, and Egypt, is approximately self-sufficient in oil at present, but their known resource base is not extensive. The others are all dependent in greater or lesser degree on imported oil.

For the non-OPEC countries as a group, excluding China, the share of mineral fuels in total imports rose from 12 percent in 1973 to 23 percent in 1977—a product of the continuing growth in consumption and the sharp increase in prices.[3] Data are not yet available on the effects of the 1979–80 price increases, but they have almost certainly made imported oil an even larger share of total imports, placing heavy strains on balances of payments. The issues of price and availability of imported oil have thus become of central concern to the prospects for continued economic growth of most of the developing world.

The Traditional Sector

Most people in developing countries are engaged in agriculture and livestock tending—growing food for themselves and for the urban minorities, often supplemented by cash crops such as coffee, sugar, or fibers. Although some agriculture is organized in plantations,

[2] Calculated from data in United Nations, *1978 Yearbook of International Trade Statistics* (New York, 1979).
[3] Ibid.

(Areas of circles are proportional to totals for each group)

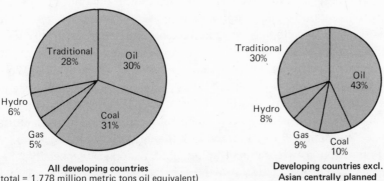

All developing countries
(total = 1,778 million metric tons oil equivalent)

**Developing countries excl.
Asian centrally planned**
(total = 1.027 million metric tons
oil equivalent)

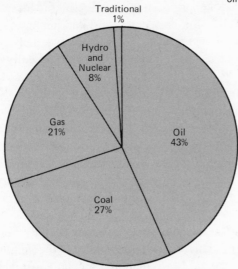

Industrial countries
(total = 5,117 million metric tons oil equivalent)

Figure 0-2. Energy Supplies in Developing and Industrial Countries, 1978. Traditional energy data for 1973 from Jyoti K. Parikh, "Energy and Development," World Bank Public Utilities Report No. PUN 43 (Washington, D.C., August 1978), recalculated on the assumption that per capita traditional energy consumption remained unchanged between 1973 and 1978. Commercial energy data for 1978, except for Taiwan, from United Nations, *World Energy Supplies, 1973–1978,* Series J, No. 22 (New York, 1979), with data on hydroelectricity and nuclear energy recalculated on basis of thermal generation primary energy equivalents. Data for Taiwan from an oral communication from the Coordination Council for North American Affairs to Lincoln Gordon. Population data from Population Reference Bureau, *1978 World Population Data Sheet* (Washington, D.C., 1979).

5

cooperatives, or modernized family farms with high productivity, the bulk continues to rely on ancient methods of production, processing, and consumption. The traditional rural sector, where most of the poverty in the developing world is concentrated, depends for its energy sources mainly on human and animal muscles, supplemented by traditional fuels for cooking, lighting, and heating where needed. Wood and charcoal are also important energy sources for household use by the urban poor. Historically, fuelwood and cattle dung—but not charcoal—have generally been "free" for gathering, and agricultural wastes are mainly burned directly by the farmers producing them.

Population growth strains traditional fuel supplies, especially when forests are displaced by farmland. In the past, as traditional fuels became more "expensive"—either because their collection required increased efforts (for example, longer daily walks to forage for firewood) or because they became partly appropriated and commercialized for cash sale—they were replaced by commercial fuels, especially for cooking and lighting, first by kerosine and subsequently by bottled gas and electricity. With the quantum jumps in oil and gas prices, these types of fuel succession have become much more costly, placing still further strain on traditional sources. Such strains may lead to vicious circles of environmental degradation: the heightened burning of cattle dung and vegetable wastes, depleting the soil of needed nutrients; and sometimes additional deforestation leading to soil erosion and desertification, fouling of downstream waters, and siltation of reservoirs. They also lead in some cases to reduced amounts of cooking and lower nutritional standards.

Traditional fuels are not suited to one especially important use in the rural sector of many developing countries: the pumping of groundwater for irrigation and for human and animal consumption. For that purpose, primitive animal-powered pumps are normally replaced by diesel engines or by electric motors when electricity is available. In transportation, likewise, the typical succession is from animals to gasoline or diesel fuel. Kerosine lamps or electricity are also preferred for lighting at a quite early stage of development. But there is extensive use of traditional fuels in industry: brick and tile making, processing of food and tobacco, sugar extraction, and even steel making in a few countries. Such uses encourage the commercialization of traditional fuels and compete for supplies with the household uses of the rural poor.

Because of the overlapping uses of traditional and commercial fuels, pressures on the supply of either one can be rapidly transmitted to the other. Since development calls for higher per capita uses of energy on top of the demands flowing from population growth, the outlook points toward increasing strains on traditional energy sources. Positive responses may take a variety of forms: more efficient use; more efficient production of such fuels as firewood and charcoal; new forms of conversion, such as biogas from animal and vegetable wastes; or replacement by novel forms of renewable energy or by commercial sources. But the expanding interest in biomass as a source of commercial energy, either for direct combustion or for conversion to alcohols and hydrocarbons, may add further pressures on basic resources of land and water. In most developing countries, it is clear that the era of traditional "free fuel" will have to give way to more systematic methods of securing energy, but those methods no longer include the option of low-cost oil.

The Modern Sector and Commercial Energy

In contrast to the traditional sector of developing countries, the modern sector has always depended on commercial fuels as its principal sources of energy. Almost every developing country has at least a small modern sector, typically including its administrative capital; its ports; and some industrial activity in mining, plantation agriculture, food processing, and manufacture of light consumer goods. The more industrialized developing countries have large urban-industrial complexes, manufacturing both consumer and capital goods and providing an array of commercial services that make their modern sectors strikingly similar to those of the industrial countries.

The rapid pace of economic development in most of the developing world since the 1950s has been accompanied by an even greater increase in use of commercial energy. From table 0-1, showing energy consumption growth rates for developing and industrial countries, it can be noted that the South has greatly outpaced the North in this respect since 1965, and especially since 1973. The vast preponderance of the increase, moreover, was in the form of oil. Between 1960 and 1978, the share of oil in the commercial energy supplies of developing countries rose from 24 to 42 percent, and when China and North Korea are excluded, from 56 to 62 percent. Natural gas is currently a significant energy source only for developing countries with substantial associated oil production, mainly members of OPEC. Hydroelectricity

Table 0-1. Growth of Commercial Energy Consumption, 1960–78
(annual percentage increase)

	1960–65	1965–73	1960–73	1973–78
World	4.2	5.1	4.7	2.5
All industrial[a]	5.0	4.7	4.9	1.5[b]
All developing[c]	−0.1	6.9	4.2	7.3[d]

Source: United Nations, World Energy Supplies, Series J, No. 19 (1950–74) and No. 22 (1973–78) (New York, 1975 and 1979).

[a] Australia, Canada, Israel, Japan, New Zealand, South Africa, United States, and Western Europe plus East European centrally planned economies (including USSR).

[b] Market economies only: 0.5 percent per year.

[c] All other countries.

[d] Market economies only: 6.8 percent per year.

also supplies only a modest share overall, but provides a substantial proportion of total electricity output and is especially important in some major countries such as Brazil and India.

The sharp increases in commercial energy consumption are natural concomitants of the changes in economic structure involved in development. At the early stages, those changes typically include the commercialization of agriculture, the introduction of industry for processing raw materials and supplying light consumer goods, the shift of labor from agricultural to industrial and urban service occupations, the growth of urban settlements, and the mechanization of transportation. Most of the activities connected with these growing sectors can only be carried out using commercial (as opposed to traditional) fuels, at least with current fuel-using techniques.

Developing Countries in the Global Energy Context

The developing countries are the largest suppliers of petroleum, which dominates the present world energy scene, even though the Soviet Union and the United States still rank first and third in production (as of 1980). Over two-thirds of the world's probable resources of oil and gas are in developing-country regions (see chapter 6). In the case of coal, however, the Soviet Union, the United States, and Australia account for over three-quarters of the total. Industrial countries also appear to hold the bulk of uranium resources.

As to consumption of energy, the developing countries now account for about 25 percent of the world's total, including traditional fuels. As shown in figure 0-3, their share of commercial consumption has risen from 9.5 percent in 1955 to 20.3 percent of a far larger total in 1978. Their 7.9 percent annual growth rate compares with 3.8 percent for the industrial countries. Even in fairly conservative growth projections for the two groups, the developing-country share is likely to exceed 30 percent by the end of the century.

It follows that effective international energy strategies cannot be

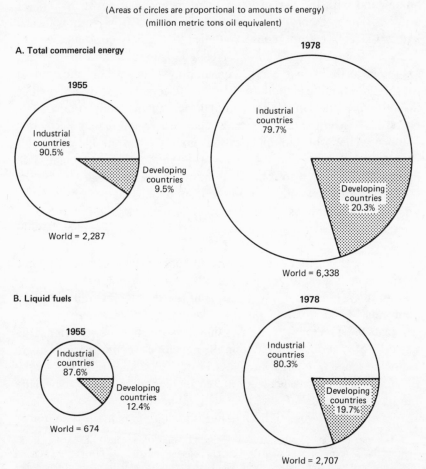

(Areas of circles are proportional to amounts of energy)
(million metric tons oil equivalent)

A. Total commercial energy

1955

Industrial countries 90.5%

Developing countries 9.5%

World = 2,287

1978

Industrial countries 79.7%

Developing countries 20.3%

World = 6,338

B. Liquid fuels

1955

Industrial countries 87.6%

Developing countries 12.4%

World = 674

1978

Industrial countries 80.3%

Developing countries 19.7%

World = 2,707

Figure 0-3. World Consumption of Commercial Energy and of Liquid Fuels, 1955 and 1978. Data are from United Nations, *World Energy Supplies, 1950–1974* and *1973–1978,* Series J, Nos. 19 and 22 (New York, 1975 and 1979).

formulated or implemented without taking the developing countries into account. There is a complex three-way interaction among development policy, energy policy, and international trade and finance.

Energy and Development

The relationship between energy use and economic development is close, but not as simple or straightforward as is often assumed. In a broad sense, social historians have virtually equated the progress of humankind with the use of animal and mechanical power to replace arduous human labor. More narrowly, it appears from both historical and cross-sectional intercountry studies that energy consumption and economic output are highly correlated. The relationship is often expressed as the ratio of energy consumption to gross domestic product (or E/GDP). But the quantitative studies are complicated by ambiguities in both concepts. On the side of GDP per capita, crude comparisons based on current exchange rates and widely circulated in United Nations and World Bank publications greatly overstate the disparities in purchasing power; in this overview, therefore, the nominal GDP figures are corrected to reflect purchasing power equivalents. On the energy side, uncertainties arise both from the absence of reliable data on the large component of traditional fuels and from the differences in how efficiently various fuels are converted to useful applications of energy.

With rough corrections for these factors having been made, table 0-2 presents a general comparison of energy consumption levels and energy intensities (or E/GDP ratios) in the developing and industrial countries.[4] The industrial countries use nine times as much energy per capita as the developing countries (line 1), while the intensity of energy use (line 5) is about twice. Within the developing-country group, there is a wide range of energy intensity. Historical studies of the industrial countries also show a phase of increasing energy intensity, especially during the periods of rapid growth in heavy industry, with later reductions of incremental E/GDP ratios to unity or below as economies

[4] Although the data in table 0-2 refer only to market economies, rough calculations indicate that, except for per capita total energy consumption, the ratios in the last column would not be significantly altered by including the European centrally planned economies in the industrial-country group and the Asian centrally planned economies in the developing-country group. When the centrally planned economies are included, the ratio of per capita total energy consumption between the two groups, as indicated in figure 0-1, is about 7.7 rather than 9.0.

Table 0-2. Energy Consumption and Gross Domestic Product in Market-Economy Industrial and Developing Countries, 1977

	Unit	Indus-trial countries	Developing countries	Ratio of industrial to developing
1. Per capita total energy consumption (incl. rough estimates for traditional fuels and counting hydropower at equivalent primary fossil fuel rates)	kilograms oil equivalent	4,698	524	9.0
2. Real GDP per capita (Kravis calculations)	U.S. dollars (1970 prices)	3,989	627	6.4
3. Energy consumption per million dollars of real GDP	metric tons oil equivalent	1,178	836	1.4
4. Assumed average fuel conversion efficiency	percentage	50	35	1.4
5. Useful energy consumption per million dollars real GDP	metric tons oil equivalent	589	292	2.0
6. Population	millions	790	1,955	0.4

Note: Because the Kravis calculations on international comparisons of real gross domestic product do not include the centrally planned economies, the data in this table (in contrast with table 0-1 and figures 0-1, 0-2, and 0-3) exclude Eastern Europe and the Soviet Union from the industrial-country category and exclude China, North Korea, and Mongolia from the developing-country category.

Sources: Traditional energy data for 1973 from Jyoti K. Parikh, "Energy and Development," World Bank Public Utilities Report No. PUN 43 (Washington, D.C., August 1978) p. 2.25, recalculated on the assumption that per capita traditional energy consumption remained unchanged between 1973 and 1977. Commercial energy consumption from United Nations, *World Energy Supplies, 1973–1978* Series J, No. 22 (New York, 1979), with data on hydroelectricity and nuclear energy recalculated on the basis of thermal generation primary energy equivalents. GDP and population data based on R. Summers, I. B. Kravis, and A. Heston, "International Comparisons of Real Product and Its Composition: 1950–1977," *Review of Income and Wealth,* Series 26, No. 1 (March 1980).

matured and their structures shifted toward services and more complex manufactured products with high added value in transformation.

Without implying that the desired course of development is necessarily in the same image as today's industrial countries, the question of whether the experience of increasing energy intensities must be duplicated by the developing countries is of great importance to their prospects. Those patterns of increasing energy intensities came about in the industrialized world at a time when such heavy industries as electric power generation and steel making were less energy efficient than they are today and when the fuel mix yielded less useful energy than today. However, since the oil price rises of 1973, energy intensities in the industrial countries have been substantially reduced, partly through public policies to encourage conservation but mainly through the response of energy consumers to altered relative costs. In effect, other factors of production, such as labor or capital, are being partly substituted for the higher priced energy. For that reason, the higher prices are not now expected to constitute as severe a constraint on economic growth rates in the industrial countries as was feared a few years ago.

For the developing countries, the outlook is more uncertain. Some considerations suggest that substitution possibilities are less extensive than in industrial countries. Moreover, the modern sectors are likely to be less flexible, with a narrower spectrum of industrial activities and fewer broadly educated managers and technicians with facility in changing products and processes. Because of the far lower living standards, there is much less inadvertent "waste." Most of the developing countries are in mild climates, with little energy devoted to space heating—an important field for conservation in colder regions. On the other hand, the smaller energy-using infrastructure of the developing countries and their more rapid rates of economic growth should give them greater opportunity in choosing technologies and strategies of development that may economize on energy. Those choices, however, may be limited by shortages of capital and foreign exchange. On balance, it seems prudent to assume fairly limited substitution possibilities, at least for the medium term.

Crisis and Transition

The world energy situation presents the oil-importing developing countries with elements of short-term crisis and long-term transition. In the short term, the problem is how to cope with sharply higher oil import bills while minimizing the adverse impact on continuing development. That calls for a combination of adjustments in other imports and exports, oil conservation and substitution, and international private or public financing of the difference. For the longer term, the problem is how to accomplish an orderly transition to an altered régime of energy supply and use involving higher relative costs, different resource demands, and possibly significant modifications in development strategies.

When the world oil price was abruptly quadrupled in 1973–74, there were widespread predictions of economic catastrophe for the oil-importing developing countries. In fact, most of those nations suffered less during the mid-1970s than the industrial countries. As a group, their economic growth rates were only slightly reduced; their energy consumption continued to expand; and their strong demand for imported goods helped to limit the depth and duration of the recession in the industrial countries.

The main keys to this unexpectedly smooth adjustment were the rapid increase in OPEC imports of both goods and services and the recycling through the commercial banking system, backstopped by the International Monetary Fund and other official institutions, of residual surpluses of "petrodollars"—dollar revenues from oil sales beyond the capacity of the oil-exporting countries to spend currently. Some developing countries received large remittances from emigrant workers in OPEC nations, while others shared in their import boom. The increase in international lending went beyond the coverage of current account deficits and thus made possible a huge increase in non-OPEC developing-country foreign exchange reserves. Meanwhile, the real price of oil was reduced from 1974 levels by virtue of the general inflation.

Can the new surge of oil price increases in 1979–80—and others perhaps still lying ahead—be managed as readily as those of the 1970s? Some observers believe that possible, but the majority are more pessimistic. The annual OPEC surplus on current account, which had been reduced from $60 billion in 1974 to $7 billion in 1978, is expected to exceed $100 billion in both 1980 and 1981, with more than half of the corresponding deficit falling on developing countries. It is unlikely that OPEC imports can again be raised so rapidly from their already high plateau. Oil-pricing policy appears to be directed much more consciously at maintaining real value plus an annual increase, if necessary by unilateral production restrictions with some degree of tacit coordination. High interest rates are increasing the annual cost of servicing external debt. The commercial banks show signs of nervousness about the creditworthiness of many developing-country borrowers, and several of those countries have begun to curtail imports of essential goods and raw materials required for development and to impose a harsh monetary austerity on their economies.

A combination of domestic and international measures will be required to surmount this crisis. OPEC countries, which have undertaken a number of new aid programs since 1974, could further cushion the effects through concessional prices or credit terms for oil-importing developing countries or through more automatic mechanisms for recycling a portion of petrodollar surpluses into developing-country investments. Cooperation among central banks and the International Monetary Fund will have to be enlarged and intensified. Debt rescheduling is likely to become more common. But insofar as the crisis reflects a basic and permanent shift in long-term energy costs, technologies, and sources of supply, and not merely the skillful manipulation of prices by a cartel, the developing countries cannot expect to be fully shielded from its effects.

On the domestic side, measures for dealing with the energy crisis in developing countries are partly matters of general macroeconomic management and partly specific to the energy sector. The latter category merges into the longer term transition—in effect accelerating and anticipating changes in energy supply and use. For most developing countries, the future energy régime will seek to be less dependent on imported oil and ultimately less dependent on oil and gas regardless of source. The more permanent energy sources are virtually certain to be more costly than oil was in the 1960s, although not necessarily more than marginal supplies of imported oil in the early 1980s. The transitional policies should therefore aim at economizing on energy— reducing the energy input needed to achieve a desired level and pattern of goods and services, and securing that energy in forms which are most efficient in terms of overall resource use.

The transition is enormously complicated by the fact that the energy supply mix of the distant future is not yet clearly defined. Transitional strategies must therefore incorporate a large component of research and development on new energy sources; they should keep options open as long as possible; and they should maintain maximum resilience to respond to unexpected opportunities or challenges. Yet the urgency of the crisis requires forward movement on replacement of imported oil and improvements in energy efficiency, accepting some risks of second-best solutions. A crucial aspect of the crisis, indeed, results from the absence in most countries of coherent policies for assuring a gradual and nondisruptive longer term transition.

These characteristics of the energy transition apply as much to industrial as to developing countries. The developing countries must also deal with a number of special problems of transition management. But it is important to avoid faddish prescriptions or generalizations about the Third World as if it were a homogeneous grouping. In particular, industrial-country spokesmen should avoid the temptation to advocate low technology and renewable energy sources as universal solutions to developing-country energy problems. Such preachments may understandably be interpreted as advice to abandon the goals of economic modernization or as disguised efforts to keep the world's dwindling hydrocarbon supplies as a preserve of the already industrialized countries.

National energy issues can be classified broadly under three headings: energy policy in relation to broader development strategies; energy conservation and improved efficiency in use; and energy supplies. In addition, the efforts of developing countries can be greatly reinforced by complementary measures of international cooperation.

Development Strategies and Energy Policies

Because energy costs and availabilities are now a major potential constraint on development, comprehensive energy planning has become an important objective for most developing countries. It is essential, however, that the energy sector not be planned in isolation from broader national objectives. It should rather be placed in its properly subordinate position, while recognizing that the changing economics of energy may entail modifications in the overall strategies.

The range of alternative development strategies widely discussed in the contemporary literature does not correspond to clearly defined energy alternatives. The categories include (a) import-substituting industrialization, (b) export-oriented industrialization, (c) balanced agriculture and industry with priority for comprehensive rural development, (d) priority for "basic human needs," and (e) radical alternatives such as "another development." The last of these five, with its severe limits on industrialization, is closely allied to proposals for exclusive reliance on "soft" energy technologies organized at the household or community level. It would probably involve substantially less per capita energy consumption than the others, but it has won few adherents among developing-country policy makers.

At the level of development tactics, on the other hand, energy considerations may be highly relevant to specific choices. Examples can be found in industrial policy, concerning the mix of products; in technology policy, concerning modes of production; and in regional policy, concerning the geographic distribution of settlement and economic activities. Rural electrification is an especially important aspect of regional policy. Energy issues are also major factors in decisions about systems of transportation and about urbanization policies. In each of these areas, plans may need to be revised in the light of altered energy costs and forms of supply.

For all developing countries now importing oil, one critical strategic issue is the extent to which they should seek self-sufficiency in energy. The pervasiveness of energy in all types of economic activity makes security of supply a cardinal aim of public policy, and nations generally prefer having vital resources under their own control. Depending on their resource endowments, however, many countries will find that self-sufficiency in energy (as in the case of food) would be so costly as to do more damage to overall development than a continuation of oil imports at a reduced level or their replacement by imported coal or other lower cost energy sources. Security against short-term supply interruptions is usually achieved more economically by stockpiling and diversification of sources. Once appropriate allowances for indirect costs and benefits are put into the accounting, the basic principle of comparative advantage still applies, and some countries will find it more rewarding to pay for imported energy by increasing exports than to eliminate their imports of energy. In the development of new forms of energy supply, moreover, international cooperation may often be more promising than the pursuit of energy autarky.

Conservation and Improved Energy Efficiency

As previously noted, their generally low levels of per capita consumption, together with the limited use of energy for space heating, make it unlikely that the developing countries can reduce their energy intensities in as large proportions as the industrial countries. There are, however, important opportunities for more efficient use of both commercial and traditional fuels.

The starting point for effective policy making in this area is the systematic collection of data on energy consumption by major forms of supply and for each of the main end-use sectors. A standardized format of energy balance sheets can help to identify substantial conservation opportunities. Although the central focus in the short term is likely to be on savings in oil, it is also important to include traditional fuels, whose more efficient use could help to reduce the damage from deforestation while relieving pressures on commercial fuels as replacements.

Conservation of Imported Oil

The transportation sector typically expands very rapidly in the course of development and in its modern forms is heavily dependent on oil. Conservation measures may be directed at less rapid rates of growth of transportation, a shift toward more energy-efficient transportation modes, or an improvement in energy efficiency of a given mode. Even at similar levels of income, developing countries show wide variations on each of these counts—for example, in number of automobiles per unit of GDP; in proportions of freight carried by road, rail, water, or air; and in efficiency of the automotive fleet. Urban settlement policies and arrangements for mass transit also bear directly on possibilities for energy conservation.

The industrial sector is generally the largest single consumer of commercial fuels in developing countries. Here too, conservation opportunities arise partly from changes in the mix of products and partly from changes in methods of production. Highly energy-intensive industries previously contemplated on the basis of cheap imported oil as feedstocks or fuels may no longer be appropriate for developing countries lacking indigenous supplies. But even for the same product, industrial processes can vary widely in their use of energy. Since most developing countries are still in the early stages of industrialization, they are often able to select energy-conserving processes from the start, rather than retrofitting or accelerating the replacement of existing equipment. They must, of course, beware of technologies replacing energy by intensive use of capital—almost by definition a scarce factor of production in developing countries. Depending on the resource endowment, many industries may also find it possible to replace oil with gas or coal, and in a few cases with biomass or direct solar energy.

Because there is little space heating in developing countries, cooking and lighting with kerosine are the principal uses of petroleum products in the residential sector. Here some gains could be achieved through improved efficiency of appliances, but the larger opportunities lie in bypassing the traditional use of kerosine in favor of charcoal, gas, or electricity generated from primary sources other than oil.

Electricity generation is the other large user of petroleum, often ranking with industry. Developing countries show a much wider range of electricity consumption intensity (in relation to GDP) than industrial countries. Economies of scale, indivisibility of large hydroelectric projects, and the capital intensity of the electric sector have traditionally led to rate structures designed to promote off-peak consumption. Where oil is the primary fuel, however, its direct costs are now tending to dominate capital costs, making it desirable to conserve electricity off-peak as well as on. That shift calls for basic changes in rate structures.

Traditional Fuels

A salient characteristic of traditional fuels such as firewood and vegetable wastes is their very low efficiency of use, generally estimated at 10 percent or less. That fact appears to open up major possibilities for conservation. Draft animals are also used very inefficiently, but the technologies for improvement are not so readily at hand. Because cooking is believed to account for some 80 percent of traditional fuel consumption in developing countries, the shift to more efficient cookers is an objective of high priority. Many types of improved cookers exist, but their adoption requires organized efforts to alter ancient habits and in many cases financial assistance to help the rural poor cover even quite small cash outlays. Efficiency gains may also come from new ways of converting biomass fuels, such as improved pyrolysis of wood to charcoal or the generation of biogas from animal and vegetable wastes.

Energy Pricing Policies

In the developing world no less than in industrial nations, the most effective single incentive to use energy more efficiently is for it to become more expensive. Because of energy's pervasive use—and in the case of electricity its status as a regulated or state-owned monopoly—its prices are subject to a wide variety of special taxes, subsidies, and controls. Even in the absence of subsidies, price increases often lag behind the increases in real replacement costs, which should be the governing criterion for pricing aimed at efficiency in energy use. The prospect of higher energy prices, however, raises serious equity problems for the household and transportation needs of lower income groups, and developing countries often lack social security or welfare systems as alternative means for handling such problems. Where subsidies are retained, it is important to consider special measures to minimize their "leakage" to higher income groups and to limit perverse effects on energy supply and use. In addition to reforms of energy pricing, moreover, complementary measures may be needed to secure conservation objectives—measures such as loans and subsidies for energy-saving investments, regulation of energy efficiency of vehicles and major appliances, and in extreme cases, some form of rationing.

Supply Alternatives

Whatever may be accomplished through the revision of development strategies and by way of conservation and improved efficiency, the largest constituent of the energy transition for developing countries will necessarily lie on the supply side. For most developing countries, there is an urgent need to find lower cost replacements for imported oil at present levels of energy consumption. And even if the historical tendency toward increasing energy intensity in the early and middle phases of development can be altered, the dynamics of population and income growth will surely call for growing energy consumption, on both a per capita basis and overall.

Domestic Mineral Resources

Fossil fuels and conventional hydroelectricity will continue to be the main commercial energy forms for at least the next two decades. Geological prospects for new discoveries of oil, gas, coal, and uranium in developing countries are sufficiently good to warrant major efforts at exploration. Even if there are no more "Mexicos" or "Saudi Arabias," smaller resource discoveries could transform the balance-of-payments and development prospects of the countries directly involved and help relieve pressure on global energy markets. Fossil fuel exploration and development in many areas have been limited by problems of finance and management, which are now being attacked through new forms of cooperation among governments, international financial institutions, and multinational corporations. In addition to conventional hydrocarbons, heavier oils, tar sands, and shale oil are coming within range of economic competitiveness.

Until recently, the ready availability of cheap oil has discouraged exploration for coal, while uranium has faced an erratically fluctuating market. Increased attention to these fuels may not only help meet domestic energy needs, but also become the basis for important new export earnings for some countries. Even though most known coal reserves are in the North, their volume is so vast that if a large and competitive world coal market comes into being, some developing countries will want to consider imported coal as a partial replacement for imported oil.

Renewable Energy Resources

Renewable resources include the biomass supplies of the traditional sector and the water power which already constitutes an important component of developing-country electricity supply. In modernized form, they are likely to play an increasing role in total energy supplies and, in the very long run, the dominant role. But there is great variety in basic endowments of renewable resources and in technical and economic suitability for large-scale commercialization.

The hydraulic resource is long established, and the increased cost of competing fossil fuels has made it much more attractive. For many countries, the order of magnitude is very large in relation to foreseeable electricity needs. Small-scale mini-hydro installations could also make a contribution, but their current vogue should not be allowed to overshadow the large remaining potential of conventional hydroelectric development.

In the few locations with natural dry steam or high temperature wet steam, the geothermal resource is already competitive. The more widespread resources of lower temperature wet steam, hot liquid, and hot dry rock may be usable in the future, but the technology is not yet satisfactory, and developing countries will be wise to await the results of costly experimentation in more affluent nations. The tropical or subtropical locations of most developing countries suggest advantageous conditions for use of direct solar energy, even though differences in cloudiness can more than offset differences in latitude. Practical applications, such as crop drying and water heating, depend heavily on specific local climate conditions and on whether or not the energy must be stored for use in nonsunny periods.

Direct and indirect uses of biomass appear to be the most promising types of renewable energy resources. Supplies could be enormously expanded by systematic forestry management. Theoretical calculations of annual increments give impressively large figures, but must be qualified by considerations of transportation costs and the competition for land among forest products, food, fibers, feed, and fuel. Nor are the ecological implications of large-scale tropical forest plantations fully understood. In many places, however, direct combustion of biomass is already a competitive means of raising steam, while pyrolysis, gasification, and conversion to methanol, ethanol, or hydrocarbons may open up major new potentials for the future.

The fuel alcohols appear to be promising alternatives to increasingly uncertain supplies of high-cost petroleum in some developing countries. They can be produced from locally available feedstocks that are often higher yielding in tropical conditions than in most industrial countries. Although methanol appears cheaper than ethanol per unit of energy content, there has not yet been a clear demonstration of wood-methanol technology at reasonable cost, and there are uncertainties concerning the environmental effects of large-scale methanol use, including its possible toxicity. It would be helpful if the research in industrial countries into methanol production from coal could be extended to include biomass feedstocks.

Ethanol is a less risky option, already heavily used in Brazil in gasohol mixtures and on a small scale as a pure motor fuel. The main limitations are relatively high cost and competition with food crops for land and other basic resources. These constraints make it important to find feedstocks less demanding of high quality land than sugarcane, as well as more economical methods of distillation.

Fuels for Electricity Supply

Changing relative costs of various fuels and of capital for alternative technologies call for reconsideration of established practices in the generation of electricity, which is vital to the modern sector of developing-country economies. The changes tend to favor hydropower and coal, as compared to oil and gas, and in some cases to favor smaller scale installations using mini-hydro or biomass and saving on transmission costs. The nuclear option is relevant mainly to the more industrialized developing countries with large electric grids and substantial engineering infrastructures but is complicated by uncertainties in the nonproliferation policies of supplier countries and by the issues of safety, spent fuel management, and waste disposal that are still unresolved in countries with advanced nuclear technology.

Ancillary Supply Considerations and "Appropriate" Technologies

Energy supply planning in developing countries must take account of a variety of factors beyond the technical and economic comparisons which are at the core of microeconomic analysis. They include limitations on the availability of specialized labor skills or other factors of production, "lumpiness" or indivisibility of certain necessary inputs, inadequacies of capital markets and credit arrangements, opportunities for external public or private financing, and broad characteristics of the existing economic infrastructure. At times, the discrepancies between national and private evaluations of costs and benefits may have to be offset by market guarantees or subsidies.

Health and environmental impacts of energy supply have received relatively little attention in developing countries, except for the problem of deforestation. It would be well to establish monitoring systems to assess these impacts at a fairly early stage, with a view to anticipating future problems and avoiding the high costs of delayed remedial measures.

Proposed changes in energy supply methods may conflict with deeply rooted cultural patterns. They may also raise issues of equity. Technological modernization is usually associated with commercialization, even in the case of localized renewable energy resources, and commercialization often tends to widen income gaps. Community supply systems may be a partial remedy, provided that the institutional framework can meet the tests of organization and management.

There is widespread current interest in the possible application, in the rural areas of developing countries, of decentralized energy technologies, sometimes referred to as "village technologies." Some are simply more efficient ways of using traditional fuels, while others (such as photovoltaic cells) require high technology for production but are easy to use in small-scale applications. Village technologies fit into the broader class of "appropriate" technologies. They are advocated partly on the ground that simple technologies have been unfairly neglected in the past and in many conditions represent more economic alternatives, and partly on more philosophical grounds, involving relationships of technology to life-styles, social organization, and environmental quality. Examples of village technologies are solar flat plate collectors for crop drying and water heating, solar cookers, solar irrigation pumps, photovoltaic cells, small windmills, mini-hydro installations, biogas, community woodlots, and improved charcoal kilns.

While economic comparisons with conventional energy supply methods are fairly straightforward, the sociopolitical considerations are more complex. These latter kinds of judgment are best left to the national and local authorities concerned. Outsiders from industrial countries usually have no special qualifications to contribute to them. The objective should be to ensure that "appropriate" technologies are genuinely appropriate in terms of the full range of social costs and benefits and value systems. Where technologies are still highly experimental, with limited promise for application in the short term, industrial countries and aid agencies should refrain from pressing their application to the developing world before the technical and economic advantages are clearly established. It is also important to recognize the institutional and cultural complexities of technological adaptation in the traditional sector and to maintain flexibility on such issues as optimum scale for biogas and other renewable energy facilities. In some circumstances, medium or even large-scale commercial ventures appear more promising than supplies at the household or small community level.

International Energy Cooperation and the Developing Countries

As with other aspects of economic development, the main burden of resource mobilization and organization for implementing energy strategies will fall on the public authorities and the private sectors of the developing countries themselves. Their actions, however, can be given vitally important financial and technical support from industrial countries and international institutions.

Financial assistance is needed to help tide over the short-term energy crisis of oil-importing developing countries. More generally, the whole array of bilateral and multilateral aid programs should be analyzed for their energy implications, including possible implicit biases toward unduly energy-intensive technologies carried over from the era of cheap oil. For the longer term transition, donor countries should expand their programs of energy assistance. Indicated lines of supporting action include assistance in (a) overall assessments of energy resources and demand patterns; (b) training and management for the energy sector; (c) exploration for mineral energy resources and appraisal of hydroelectric potential; (d) applying modern forestry practices, including fuels as end products; and (e) transferring research results on alchohol fuels, synthetic oil and gas, and novel renewable energy technologies as and when they become viable supply sources. In the nuclear energy field, supplier-country policies need to be clarified with respect to conditions governing exports of fuels and enrichment services, spent fuel management, and requirements for international safeguards.

In broader perspective, there is a basic mutuality of interest among industrial and developing countries in a peaceful transition to a more durable energy regime sufficient to the needs of all. The oil exporters are not exceptions. Their sources of supply and markets require some degree of world economic stability, and they must consider their own ultimate transition to economic diversification and substitute energy sources. The alternative of competitive scrambles, bilateral "special deals," and threats or use of military force to secure supplies is extremely hazardous, even to those who suppose that they might secure a temporary gain.

The developing-country energy importers have not yet organized themselves for the pursuit of a joint energy strategy, notwithstanding the crucial importance of energy to their continuing development prospects and their increasing weight in the global energy balance. They would be essential participants in any systematic efforts at international negotiation of oil prices and production, energy conservation targets, and accelerated development of alternative supplies. There are formidable obstacles to effective energy cooperation on a global basis, but the stakes are so large that it would be unwise to leave unexplored any possible paths to that end.

1

The Developing Countries
and the World Energy Crisis

The global energy crisis is profoundly affecting the developing countries, which contain three-quarters of the world's population. The ways
in which that crisis comes to be resolved will greatly influence the
economic and perhaps political future of the developing countries. At
the same time, the nature of their individual and collective responses
will greatly affect the terms of the global outcome.

Four-fifths of current world energy exports come from developing
countries. Their share in global consumption of commercial fuels—
already 20 percent—is also rising rapidly. Any measures on their part
to control production, to develop new energy sources, or to improve
energy efficiency affect the tightness or softness of world oil markets.
For the developing-country energy importers, increasing costs are
likely to retard economic growth and impose severe limits on non-oil
imports, shrinking the export markets of the industrial countries and
threatening the stability of the international trading and financial
systems. But the repercussions extend beyond the economic sphere.
Although development does not guarantee domestic political stability,
sporadic or frustrated development is a certain producer of instability.
In a world already riven by ideological rivalries, scrambling for vital
energy supplies can undermine alliances and jeopardize the prospects
for international security and the basic values of freedom as well as
welfare.

The interest of the industrial world in the production and pricing
policies of the oil-exporting developing countries—especially the members of the Organization of Petroleum Exporting Countries (OPEC)—
is too obvious to need comment. Although less direct, the industrial
world also has an important stake in the successful management of the
energy problems of those developing countries that import oil. That

stake is not limited to a humanitarian concern for the reduction of poverty and misery or a political concern for the stability and welfare of old or new allies.

Even more than for most commodities, all forms of energy, traditional as well as commercial, have overlapping uses and are potential substitutes at the margin. The world oil trade is the ultimate balancer of supply and demand. A barrel saved or replaced anywhere means lessened pressure on a sensitive global equilibrium. But if the barrel is saved at the cost of economic growth, either in the developing or in the industrial countries, or if higher prices are offset by national borrowing beyond the range of prudence, other aspects of international trade and finance are placed in jeopardy. At the same time, the developing countries—including energy exporters as well as importers—have a vital stake in successful resolution of the energy problems of the industrial world, which provides the main markets for developing-country exports of raw materials and manufactured goods, the main sources of capital and technology essential to development, and most of the resources for international aid. There exists, therefore, an underlying convergence of interests that should make possible an international energy strategy with some degree of benefit for all the major parties involved.

The Short-Term Crisis and the Long-Term Transition

The term "crisis" was widely applied to the international oil developments of late 1973 and early 1974, because of the fourfold increase in export prices decreed by OPEC members and the concurrent attempt on the part of Arab OPEC members to impose a selective embargo on supplies to the United States and the Netherlands and sales restrictions on many other countries. The embargo did not succeed, but the attempt created near panic in many importing countries, concerned that their own supply lines might someday be cut off. During the mid-1970s, as energy consumers began to adjust to higher prices, the fears of monetary catastrophe proved unfounded, and oil prices were gradually reduced in real terms by virtue of general inflation. Some criticized the word "crisis" on the ground that it implied a short-term emergency susceptible to rapid resolution, whereas the most important issues concerned long-term problems of transition to an energy régime different from that of the postwar quarter century.[1] In view of the political events of 1978–80 in Iran, the related surge in world oil prices in 1979–80, and the renewed evidence of political and strategic instability in the Middle East generally, it may be fair to describe the world energy situation at the beginning of the 1980s as combining both long-term problems and short-term crisis—the essence of the crisis arising from the absence of coherent policies for ensuring a gradual and nondisruptive longer term transition.

The short-term supply crisis calls for policy responses in several areas: national and international macroeconomic adjustment to sudden price increases, including the recycling of surplus petrodollars; minimizing and managing the disruptions arising from any sharp cut in production or abrupt price rise in the oil markets (sometimes described as an oil "crunch"); and coping with the international security dangers of supply interruptions and of nuclear proliferation.

The developing countries as a group were less severely affected by the macroeconomic impact of the 1973 price increases than was feared, although critical problems arose for some individual countries. There is good reason to expect a sharper impact from the 1979–80 round. That prospect gives added urgency to the search for economically viable replacements for imported oil, as well as for means of improving energy consumption efficiencies. It also gives the developing countries a vital interest in the health of the international financial structure, including measures by the International Monetary Fund to stand behind petrodollar recycling[2] by commercial banks and any special aid provisions to assist in financing oil imports.

As to possible disruptions in the flow of supplies, OPEC ministers have given assurances of preferential treatment for developing countries, but there is no organized system to implement such preferences. Stockpiling is the normal first line of defense against supply disruption,

---

[1] See, for example, Charles J. Hitch's foreword to Sam H. Schurr, Joel Darmstadter, Harry Perry, William Ramsay, and Milton Russell, *Energy in America's Future: The Choices Before Us* (Baltimore, Johns Hopkins University Press for Resources for the Future, 1979).

[2] "Petrodollar recycling" is a term devised during the 1970s to describe the flow back to oil-importing countries of dollar receipts for the sale of oil (mainly by OPEC members) beyond the capacity of the selling countries to spend on current imports of goods and services. Most of the surplus petrodollars have been deposited in commercial banks in the West (particularly in "Eurodollar" accounts), forming the basis of short-to-medium-term loans by those banks to both industrial and developing countries experiencing deficits in their current balances of payments.

but for many developing countries this is not an easy option, either financially or physically. The other security issue—nuclear proliferation—is relevant to only a small number of developing countries, but for them the working out of international arrangements reconciling nuclear supply and nonproliferation concerns may critically affect their electricity supply prospects.

For the longer term transition, international financial issues and macroeconomic effects of higher costs on balances of payments and growth rates are also important. But the key long-term problem concerns the replacement of cheap oil and gas—partly by finding new sources of supply and new kinds of energy resources, and partly by increasing energy efficiency so as to reduce the energy inputs needed for otherwise achievable levels of economic output. This longer term transition, as it involves the developing countries, is the central problem examined in this book.

Adequate supplies of energy at manageable costs are essential to avoid stagnation or even severe economic retrogression in both the developing world and the industrial nations. The stakes involved in a successful transition are huge, but the obstacles to success are also very great. New sources of conventional fuels and new energy technologies tend to be costly. Uncertainties are inescapable, as in prospecting for fossil fuels or in estimating the ultimate costs of commercialization of energy technologies now in their infancy. The effects of political upheavals on levels of energy production and export are even less predictable. Cooperative action in worldwide organizations or among groups of energy suppliers or consumers could help to smooth the transition, but such cooperation may not prove feasible. Some aspects of energy policy—especially supply systems with heavy capital requirements and long construction lead times, or conservation measures working through the progressive turnover of a large capital stock—require firm adherence for years to achieve substantial results.

Many problems involved in the long-term transition are common to both the industrial and the developing countries. But finding adequate solutions is likely to be much harder for the latter. Even in the era of cheap oil, two of the main limitations on economic growth in the developing countries were the capacity to mobilize capital for investment and the capacity to import commodities and equipment not produced domestically. Problems of capital supply and inadequate foreign purchasing power are likely to become aggravated, since most of the promising substitute energy sources are capital intensive and many of them will require imported technology, machinery, and operating supplies. The higher prices of imported oil are already putting great strains on developing-country balances of payments, and the OPEC countries have thus far made only small moves in the direction of special price concessions or large-scale programs of development aid.

With generally low levels of per capita energy consumption, and with much less use for space heating, the developing countries cannot expect gains from conservation of commercial energy in the same magnitudes as the industrial countries. There are, however, important opportunities for more efficient use of both traditional and commercial fuels—in cooking, transportation, and some industrial and commercial applications. Newly industrializing countries can also develop their infrastructure, modes of settlement and transportation, building designs, and industrial technologies in ways that conserve energy at the outset, rather than copying patterns evolved in industrial countries when energy was cheap. At the same time, since the essence of economic development is increased productivity and structural change—in the early stages in an energy-intensive direction—the expansion of supplies is bound to constitute the main thrust of the energy strategies of developing countries. The problem is compounded by the sharp differences in birthrates, requiring substantial economic growth in the developing countries simply to provide jobs for an expanding labor force and to keep average consumption levels from falling even lower.

As to longer term supply options, the developing countries share in all the problems and possibilities of the industrial countries, but in addition confront the special problems of significant dependence on traditional energy sources. Most technologies aimed at managing these "renewable" supplies to keep them truly renewable (or at replacing them with other fuels) involve substantial capital expenditures, environmental problems, and difficult changes in institutions and cultural patterns. For certain types of renewable resources, climate and terrain may be especially advantageous in some developing countries; for others, the lack of advanced infrastructure and technological skills may be a serious disadvantage.

The extensive use of traditional fuels—mainly firewood but also including animal and vegetable wastes—contains elements of paradox. These local renewable energy sources in their traditional forms are not capital intensive, and they do not require imported technology. But they are being pressed by growing populations and by the very processes of development, which require not only more energy but also more land for agriculture or more intensive cultivation of existing farmland. The burning of cattle dung sacrifices its use as fertilizer. As prices rise, even urban dwellers can no longer afford kerosine, so they too often turn back to firewood or charcoal as cooking fuels. The vicious cycles of deforestation, soil erosion, and downstream siltation lead to further reductions in yields of both food crops and forest products. The squeeze on fuelwood comes at a time when energy planners all over the world are exploring possible new uses for wood and other biomass resources—as a base for synthetic liquid fuels to replace oil in transportation and for direct combustion in industry. These interwoven issues of land-use competition and environmental protection, combining traditional and novel energy sources and technologies, have little parallel in the industrial world.

Energy sector planning has become a necessity for developing countries, since energy costs or supply inadequacies may jeopardize the entire prospect of continuing development. Even this brief discussion will have suggested the complexities of effective energy planning—complexities of a political character as well as technical and economic. As the subsequent chapters will demonstrate, it is not now possible to identify any simple panaceas in the form of altered development strategies, massive conservation opportunities, or magical supply alternatives that will resolve this mosaic of problems. A complex of responses is called for, not a single all-embracing remedy. Every developing country faces major issues in working out patterns of energy supply and use compatible with its basic resources and its development aspirations.

The Developing Countries—Characteristics and Diversity

The term "developing countries" does not have a precise definition, but generally refers to countries with relatively low per capita incomes or with economic structures typical of the early stages of modernization, such as occupational concentration in agriculture and livestock tending, little urbanization, and low educational levels.[3] When data are assembled on the basis of geographical regions, the developing countries

[3] For a fuller discussion, see the appendix to this chapter.

comprise the three great continental areas of Africa, Asia, and Latin America. The contrasting industrial-country group consists of the United States and Canada, Western and Eastern Europe (including the USSR), Japan, Australia, and New Zealand. For obvious reasons, the two groups are often referred to as the South and the North.

In 1964, almost all the countries then constituting the South organized themselves for negotiations during the first United Nations Conference on Trade and Development (UNCTAD) into the Group of 77 (G-77). That term is still used, although the number of member developing countries is now about 127, the major nonparticipant in the group being The People's Republic of China. The group has no headquarters or secretariat, but its representatives meet as a caucus at all United Nations conferences dealing with economic issues and try to arrive at agreed positions. Partly because of these arrangements, the developing countries are often treated in the mass media as if they were a grouping with homogeneous economic, social, and political characteristics—referred to as the Third World.

A bipolar distribution between North and South, with only a few borderline cases, does in fact exist for many important variables: per capita incomes, population growth rates, occupational distribution, urbanization, literacy, and others. But the most striking aspect of recent economic history in the developing regions is their increasing differentiation, with the emergence of a class of newly industrializing countries (NICs) as the most dynamic actors in the world economy. The OPEC members themselves constitute another distinctive group. Among the continental regions, development is substantially more advanced in Latin America and East Asia than in Africa and South Asia.

In the energy sector, there are major differences among groups of developing countries in patterns of supply and use. A preliminary division can be made between oil-exporting and oil-importing developing countries. Most of the exporters are members of OPEC, the principal exception being Mexico. They range in nominal per capita gross national product (GNP) from $300 for Indonesia to over $14,000 for the United Arab Emirates, a level twice the average of the industrial countries. The more populous ones, such as Indonesia, Nigeria, Algeria, Iran, Iraq, Venezuela, and Mexico, are able to spend all their oil income on current imports, while those with small populations are the generators of the "petrodollar surpluses," which are recycled as capital exports. This study does not include the special energy issues of economies overwhelmingly dominated by oil production and exports.[4]

The larger group of oil-importing developing countries is even more diversified than the exporters. Nominal levels of output are as low as $100 to $150 per capita for most of South Asia and much of subSaharan Africa, but over $1,500 for the most prosperous countries in Latin America, East Asia, and southern Europe.[5]

Within the higher income category, two groups stand out as exceptional: (a) those in southern Europe, plus Israel and South Africa, with a relatively high level of human capital and physical infrastructure, which are often classified as industrial countries, and (b) the NICs, rapidly industrializing and expanding their exports of manufactured goods. The latter include Korea, Singapore, Taiwan, and Hong Kong in East Asia, and Brazil, Mexico, and Argentina in Latin America. Within the lower income group, it is also important to differentiate between countries like India and Pakistan, with substantial modern industrial sectors and large cadres of skilled elites, and those where almost the entire populations are engaged in subsistence farming and herding, for example, the Sahelian belt of Africa.

The People's Republic of China, with a third of the developing-country population and a quarter of the entire world's, is obviously of great ultimate importance to global energy prospects. For the near future, it appears likely that China will be only a small net exporter of fossil fuels, with modest effect on world markets. As to domestic energy uses, the available information is limited and often not comparable with other countries. For these reasons, although estimates for China are included in our aggregate figures for developing countries, this study does not deal with the more specific aspects of China's energy resources and policies.

[4] Oil exporters will also have to participate in due course in the global transition to a different combination of energy sources. Their time schedules depend on known reserves and possibly undiscovered resources, with some much better endowed than others. Some have virtually no economic resource base other than oil and face formidable problems in contemplating the long-run future. There are also challenging issues about how best to use oil revenues to achieve economic diversification and balanced growth, avoiding the tensions of a "dual economy." Those problems of coping with temporary energy affluence, however, would require a separate major study.

[5] The World Bank uses a classification of "low-income" countries with GNP up to $300 per capita (forty-three countries; total population in 1977 of 1,262 million); "middle-income" countries above $300 per capita (seventy-six countries; total population of 919 million); and "centrally planned" developing countries, including China (five countries; total population of 916 million). Vietnam, Laos, and Cambodia are included among the low-income countries, rather than the centrally planned. The World Bank, *World Development Report, 1979* (Washington, D.C., 1979) pp. 126–127.

There are many typologies of developing countries, depending on the purpose of the classification. For example, one energy study focuses on structural characteristics, grouping developing countries into (a) industrialized, (b) balanced growth, (c) mineral exporters, (d) agricultural exporters, and (e) other agricultural.[6] Other relevant factors include country size and extent of water power, fossil fuel resources, and biomass potential.

In this book, our discussion is often documented by reference to selected developing countries, drawn from these broad categories and representative in three aspects: geographical location; a range of income levels; and a variety of economic structures. The selection has also been constrained by data availability.

Energy Use in Developing and Industrial Countries

Both historical and cross-sectional intercountry studies show a strong association between energy use and other indicators of economic development. The spectrum of income levels and economic structures in the developing countries is matched by a spectrum of energy supply and use patterns, those of the higher income nations resembling the industrial countries. For the two groups as a whole, however, there are substantial differences that must be taken into account in analyzing their energy problems and policies.

Consumption Levels

Per capita energy consumption, including both commercial forms such as oil and hydroelectric power and traditional such as firewood, is nine times higher (at 4,698 kilograms oil equivalent per capita) in industrial than in developing countries (about 524 kilograms oil equivalent [see table 0-2]).[7] Within the developing world, there is considerable diversity. As shown in figure 0-1, consumption levels in Asia are about one-half of those in Latin America, while African levels are slightly lower than Asian. There is even greater variation within these broad geographical areas. Table 1-1, which deals only with commercial energy, shows a ninefold spread in per capita consumption among the

[6] See Philip F. Palmedo and coauthors, *Energy Needs, Uses, and Resources in Developing Countries* (Upton, N.Y., Brookhaven National Laboratory for the Agency for International Development, 1978).

[7] See figure 0-1 for sources.

Table 1-1. Twenty Largest Per Capita Consumers of Commercial Energy Among Developing Countries, 1978

Countries	Per capita consumption (kilograms oil equivalent)	Aggregate commercial energy consumption (thousand metric tons oil equivalent)	Population (millions)	Liquid fuel consumption (thousand metric tons oil equivalent)
Twenty largest per capita consumers:[a]				
1. Kuwait	4,606	5,518	1.2	1,784
2. Trinidad	3,377	3,827	1.1	1,602
3. Venezuela	2,200	28,872	13.1	13,095
4. North Korea	2,015	34,386	17.1	1,256
5. Singapore	1,674	3,907	2.3	3,901
6. Taiwan	1,437	24,284	16.9	17,000[b]
7. Argentina	1,325	34,962	26.4	23,357
8. Libya	1,285	3,183	2.5	3,183
9. Iran	1,250	43,989	35.2	23,174
10. Hong Kong	1,127	5,192	4.6	5,180
11. Mexico	996	66,710	67.0	42,996
12. South Korea	947	35,030	37.0	20,891
13. Saudi Arabia	888	8,718	9.8	8,101
14. Cuba	795	8,036	10.1	7,903
15. Chile	784	8,519	10.9	5,058
16. Syria	696	5,627	8.1	4,692
17. Brazil	694	80,092	115.5	46,702
18. China	582	533,197	916.2	91,917
19. Turkey	578	24,970	43.1	17,227
20. Malaysia	501	6,611	13.2	5,668
Total	715	965,630	1,351.3	344,687
Total, all developing countries	417	1,288,116	3,089.6	536,527
Percentage share of 20 in total	—	75.0	43.7	64.1
For comparison:				
United States	8,111	1,784,312	220.0	801,199
United Kingdom	3,678	205,309	55.8	79,726
France	3,297	175,689	53.3	99,480
Japan	2,763	317,487	114.9	222,197

Note: Dashes = not applicable.

Sources: United Nations, *World Energy Supplies, 1973–1978*, Series J, No. 22 (New York, 1979), with hydroelectricity and nuclear energy converted to thermal generation primary energy equivalents. Figures for Taiwan are from oral communication with the Coordination Council for North American Affairs.

[a] Limited to countries with population greater than 1 million and with total consumption greater than 3 million tons oil equivalent.

[b] Rough estimate.

twenty highest consuming countries, with the top of the range approaching the industrial-country norm.

These great disparities in energy consumption per capita are primarily related—both as cause and as effect—to differences in levels of income (or gross domestic product [GDP]). Levels of energy consumption per unit of gross domestic product are much closer (see table 0-2, lines 3 and 5).[8] Even so, industrial countries are shown to consume about 40 percent more energy than the developing countries, relative to income.

Energy Sources and Import Dependence

In industrial countries, virtually all energy consumed comes from commercial sources—coal, oil, gas, and electricity produced from fossil fuels, water power, and nuclear energy. Oil is the major single fuel, supplying almost one-half of total consumption.

In contrast, the developing countries consume more than one-quarter of their total energy in the form of traditional fuels such as wood, charcoal, animal dung, and crop wastes (see figure 0-2). This proportion varies considerably among regions, reaching over 60 percent in Africa, 34 percent in South and Southeast Asia, and about 25 percent in Latin America.[9] Fuelwood is the principal source of household energy in most developing countries, particularly in rural areas. Oil is important in all regions, but gas is little used except in Latin America and the Middle East. Coal is heavily used in only a few cases where mines were developed before the era of cheap oil, notably China, India, Zambia, Zimbabwe, and North Korea. The contribution of electricity in the least developed countries is small at present, but in some uses, such as powering irrigation pumps and industrial motors, it is considered critical to development.

[8] GDP is converted to U.S. dollars in this table by using purchasing power parity rather than the market rates of exchange used in United Nations and World Bank comparisons of nominal GNP per capita. The conversions here are based on the actual purchasing power of the income in the countries in question, which is not accurately reflected by official exchange rates. The purchasing power parities come from Irving B. Kravis, Alan W. Heston, and Robert Summers, "Real GDP Per Capita for More Than One Hundred Countries," *Economic Journal* (London) vol. 88 (June 1978) pp. 215–242. The effect of using purchasing power parity rates of exchange rather than market rates is to increase the apparent level of real income of developing countries relative to industrial countries, thus reducing their energy/GDP ratios.

[9] As we point out in chapter 2, systematic data on consumption of traditional fuels are not compiled by national authorities. The figures here should therefore be regarded only as representing very rough orders of magnitude.

As a group, the industrial countries depend on imports from developing countries for over one-third of their supplies. In 1978, individual countries were much more heavily dependent; Japan imported 90 percent of its total energy supplies and France 78 percent. On the other hand, some industrial countries have relatively low levels of import dependence, for example, the United States at about 25 percent. The only net exporters in this group are Canada, Norway, and the Soviet Union, soon to be joined by the United Kingdom.

Because OPEC members are counted in the developing-country category, this group as a whole is a net energy exporter. Excluding the OPEC members, developing countries are generally somewhat less import dependent than industrial countries because the traditional fuels are always locally supplied. For the majority of the developing countries, however, imported oil is a large component of their total energy supplies, reaching very high proportions in such cases as Korea, Brazil, and Jamaica.

Efficiency of Fuel Supplies

What matters to energy consumers is not the gross amount of energy used, but the energy services received (or "useful" energy). Traditional fuels are typically used in ways that yield very low efficiencies. Thus fuelwood is mainly used for cooking over open fires, in which only 10 to 15 percent of the gross input is received in the form of useful energy.

Commercial fuels generally yield higher efficiencies. It follows that countries using primarily commercial fuels (the industrial group) will receive more energy services for a given energy input than those (the developing countries) which depend largely on low-efficiency traditional fuels.[10] Making a rough adjustment for these differing average efficiencies (table 0-2, line 4), it appears that the industrial countries consume about twice as much useful energy, relative to income, as the developing countries.

---

[10] An attempt to take account of differences in average efficiencies in calculating energy intensities is made in table 0-2. The average end-use efficiency of commercial fuels is assumed to be 50 percent in both industrial and developing countries, and an average efficiency of noncommercial fuels, applying only to developing countries, is assumed to be 10 percent. Together these yield an average efficiency of the overall fuel supply of 35 percent in developing countries and 50 percent in industrial countries (line 4).

Energy Embodied in Imports and Exports

A nation's energy consumption does not depend entirely on the amounts of energy consumed directly. Energy is also consumed indirectly as part of imports of a wide range of goods and services. Similarly, some of the energy counted as domestic consumption will be used up in the manufacture of goods for export. If a country imported all its requirements of energy-intensive goods—say iron and steel—and exported only less energy-intensive goods—agricultural products, for example—then its domestic energy consumption (measured by production plus imports minus exports of energy) would be lower than if it imported agricultural goods and exported iron and steel products.

An analysis of the indirect or embodied energy contained in the foreign trade of nine industrial countries (also taking into account that part of petroleum refinery consumption used to provide net exports of refinery products) indicates that the United States, Canada, France, and the United Kingdom had small net imports of embodied energy, while Germany, Italy, the Netherlands, Sweden, and Japan were substantial net exporters.[11] A similar study of the trade of oil-importing developing countries shows them to be significant net importers. The size of the net import (measured as a percentage of total energy consumption) varies considerably from country to country, but is rarely less than 10 percent and more often around 30 percent. Those countries with the lowest percentages (India, Korea, Mexico, Argentina, and Brazil) are the most industrialized of the developing countries.[12]

In general, therefore, the developing countries are net importers of embodied energy and the industrial countries net exporters. Part of the difference in energy consumption between the two areas results from the nature of their trading relationships.

Differences in the Structure of Energy Consumption

There are also major differences in the structure of energy demand between industrial and developing countries. It is customary to identify

[11] Joel Darmstadter, Joy Dunkerley, and Jack Alterman, *How Industrial Societies Use Energy* (Baltimore, Johns Hopkins University Press for Resources for the Future, 1979).

[12] Alan Strout, "The Hidden World Trade in Energy," unpublished paper Massachusetts Institute of Technology Energy Laboratory, Cambridge, Mass., 1975.

Table 1-2. Sectoral Energy Consumption as a Percentage of Total Energy Consumption (Including Noncommercial Fuels) in Selected Countries, 1976
(percentage)

	Industry[a]	Transportation	Other[b]	Electricity as percentage of energy consumption
OECD	39	27	34	13
OECD (Europe)	43	20	37	13
U.S.	34	33	34	13
Japan	60	16	24	15
Brazil[c]	34	25	39	9
India[c]	43	17	39	7
Kenya	9	15	76	3
Mexico[c]	51	27	14	7
Nigeria[c]	8	9	76	1
Portugal	46	36	18	14
Turkey	23	24	53	5

Sources: Organisation for Economic Co-operation and Development, Energy Balances of OECD Countries—1975–77 (Paris, 1979); International Energy Agency, Workshop on Energy Data of Developing Countries (Paris, Organisation for Economic Co-operation and Development, 1979). Data include traditional as well as commercial fuels.

[a] Includes fuel and electricity consumption of industry, feedstocks in the chemical and petrochemical industries, and all nonenergy uses of fossil fuel materials.

[b] Mainly household and commercial.

[c] For these countries, totals under 100 percent are explained by energy use unaccounted for.

three major energy-consuming sectors: industry (including energy materials used as feedstock for petrochemicals), transportation, and "other" (which includes agriculture, household, commercial organizations and handicrafts, public services, and miscellaneous uses).

As shown in table 1-2, the two largest sectors in the OECD countries are industry and "other" (mainly residential), accounting for about 40 and 30 percent respectively. Transportation comes next with about 20 percent. Within this group of countries, however, there are many variations. The share of industry is much higher in Japan than in the United States, and the share of transportation is much lower in both Japan and OECD Europe than in the United States.

While comparable information is not available for all developing countries, the data for selected countries given in table 1-2 permit some rough generalizations. It will be seen that the newly industrializing countries show levels and patterns of energy use closer to the OECD group than to the poorer developing countries. Among the poorest countries, consumption for household purposes is dominant. In part this results from the inefficiency in use of fuelwood referred to earlier, but it also reflects the much greater relative importance of the rural sector. As countries grow richer, their patterns of energy consumption change; the household share diminishes while industrial consumption grows.

The share of electricity is also highly sensitive to level of development. In the poorest countries electricity accounts for only 1 to 3 percent of final energy consumption. As development proceeds, the percentage rises, reaching an average of about 13 percent for industrial countries. The rising share of electricity accounts for part of the higher total energy consumption of the industrial countries, since about two-thirds of the primary fuel energy content is lost in heat in the process of electrical generation.

The Global Energy Framework and the Developing Countries

The developing countries are the largest suppliers of petroleum, which dominates the present world energy scene, even though the Soviet Union and the United States still rank first and third in production. On the basis of presently known reserves and probable recoverable resources, the developing-country predominance in supply of both oil and natural gas will continue for the indefinite future. One recent attempt to consolidate estimates for all depletable fuels shows developing-country regions with over two-thirds of total probable resources of these hydrocarbons.[13] That is in striking contrast with coal, where the Soviet Union, the United States, and Australia account for over three-quarters of the total, with only China well endowed among developing countries. Much less is known about uranium, but there, too, North America and Australia appear to possess the largest resources.

Another noteworthy feature of oil is the very large fraction of total supplies which moves in international trade. In 1978, no less than 1,705 million metric tons (or about 34 million barrels per day [mbd]) were exported, some 57 percent of the total production of 2,986 million tons (60 mbd). In contrast—and notwithstanding the importance of the world grain trade—the total exports of food and feed grains together

[13] Calculated from tabulation in Hans H. Landsberg and coauthors, *Energy: The Next Twenty Years*, a study sponsored by the Ford Foundation and administered by Resources for the Future (Cambridge, Mass., Ballinger, 1979) pp. 246–247.

amounted to only 11 percent of production. Trade in other fuels was also relatively small. Of the total exports of oil, 80 percent came from OPEC member countries, 6 percent from Mexico and other developing countries, and only 14 percent from industrial countries (mainly the USSR, Canada, Norway, and Britain).[14]

On the demand side, the developing countries account for an already large and rapidly growing share of total commercial energy consumption. (If traditional fuels were included, their share would be substantially larger.) Figure 0-3 graphically displays three salient aspects of change in the world's commercial energy economy from 1955 to 1978. They are: (a) the large increase in total consumption (2.8 times, or 4.5 percent per year); (b) the even greater increase in liquid fuels (6.2 percent per year), bringing them from 29 to 43 percent of total commercial energy; and (c) the spectacular growth in the developing-country share (from 9.5 to 20.3 percent of the total, reflecting a 7.9 percent annual growth rate).

Although there are about 127 developing countries, the analysis of energy issues—especially those related to global supply and demand—can often be confined to a much smaller number because of the high degree of concentration. Table 1-3 shows the data for the twenty largest consuming countries, which account for 86 percent of the commercial energy consumption of all developing countries (the ten largest consume 75 percent). Six of these countries are oil exporters, whose high consumption is related to the energy intensity of the petroleum production and refining industries. The two countries heading the group—China and India—are there by virtue of sheer size. The list also includes almost all the developing countries whose electrical grids are sufficiently large to bring nuclear energy into serious consideration.

Most energy projections for the rest of the century assume that consumption growth rates for commercial fuels in developing countries will exceed those of industrial countries by 2 to 4 percentage points. Between 1955 and 1978, the difference exceeded 4 points—7.9 percent versus 3.8 percent. The reasons include higher population growth, more rapid overall economic growth, structural changes involving greater energy intensities, and replacement of traditional by commercial energy sources. If a differential of at least 2½ points is maintained, the developing-country share in world commercial energy consumption

[14] Calculated from data in table 10 of United Nations, *World Energy Supplies, 1973–1978*, Series J, No. 22 (New York, 1979).

Table 1-3. Twenty Largest Consumers of Commercial Energy Among Developing Countries, 1978

Countries	Commercial energy consumption (thousand metric tons oil equivalent)	Per capita consumption (kilograms oil equivalent)	Population (millions)
1. China	533,197	582	916.2
2. India	84,583	132	638.6
3. Brazil	80,092	694	115.5
4. Mexico	66,710	996	67.0
5. Iran	43,989	1,250	35.2
6. South Korea	35,030	947	37.0
7. Argentina	34,962	1,325	26.4
8. North Korea	34,386	2,015	17.1
9. Venezuela	28,872	2,200	13.1
10. Indonesia	28,157	192	147.0
11. Turkey	24,970	578	43.1
12. Taiwan	24,284	1,437	16.9
13. Colombia	14,141	546	25.9
14. Egypt	14,062	355	39.6
15. Philippines	11,592	250	46.3
16. Thailand	10,830	240	45.2
17. Pakistan	9,936	129	76.8
18. Algeria	8,744	473	18.5
19. Saudi Arabia	8,718	888	9.8
20. Cuba	8,036	795	10.1
Total, 10 largest	966,842	480	2,013.1
Total, 20 largest	1,105,291	471	2,345.3
Total, all developing countries	1,288,116	417	3,089.6
For reference:			
Percentage share of 10 in total	75.1	—	65.2
Percentage share of 20 in total	85.8	—	75.9

Notes: Includes developing countries with market economies, centrally planned economies in Asia, and Taiwan. Excludes southern Europe and South Africa, classified by the United Nations as "developed market economies." Dashes = not applicable.

Sources: United Nations, *World Energy Supplies, 1973–1978,* Series J, No. 22 (New York, 1979), with hydroelectricity and nuclear energy converted to thermal generation primary energy equivalents. Figures for Taiwan are from oral communication with the Coordination Council for North American Affairs.

will rise to about 25 percent by 1990 and over 30 percent by the end of the century. Those shares are sufficiently large to have a major impact on world market conditions.

It follows that international energy strategies cannot be formulated or implemented without taking the developing countries into account.

The international impact on energy supply and demand is so powerful, moreover, that for countries with any substantial consumption, realistic national energy strategies must include a major international dimension. Insofar as international strategies involve multilateral negotiation, there is a compelling case for participation by developing countries on both the energy-importing and energy-exporting sides.

Even with the focus on oil-importing developing countries, the group is so large and variegated that each country will confront some energy problems related to its particular resource endowment, geography, and development dynamics. But there are also major common problems, some also shared by the industrial world and some affecting only the developing countries. There are common elements in a rational approach to energy assessment and policy making, even with respect to issues specific to particular countries. This book concentrates on these common approaches and widely shared problems. It seeks to illuminate them in ways helpful to developing-area energy planners and to those in industrial countries and international organizations concerned with the interwoven objectives of maintaining economic development and achieving a successful long-term energy transition for the world as a whole.

In the broader perspective of relations between the industrial and developing countries, the energy sector occupies a special place. Through bilateral aid programs and a variety of international institutions, the North has for several decades been expressing an affirmative interest in the development of the South, even though the level of support has been inadequate from the southern viewpoint. But whatever the nature of that northern interest—geopolitical, economic, or humanitarian—the relationship has been an asymmetrical one of donors and recipients—of the North helping the South to become "more like us." In the energy sector, in contrast, North and South are embarked on a common enterprise, in which each country's successes contribute directly to success for all. In this quest for combinations of energy supply and use sufficient to the needs of all, and for a peaceful transition from the short-lived era of cheap oil to a more durable energy economy, the world is united by a single vast community of interest.

APPENDIX 1-A
THE DEFINITION OF "DEVELOPING COUNTRIES"

Developing countries have 75 percent of the world's population but account for only 22 percent of global economic production. This latter percentage is calculated from the translation of national accounts data for 1977 into dollar equivalents at current exchange rates, using the figures from the World Bank's widely circulated 1979 *World Bank Atlas* (Washington, D.C., 1979). International comparisons of real purchasing power have established that this practice substantially understates—perhaps by a factor of two—the production and incomes of the poorer countries, but corrected data are not available for all countries or for extended periods of time. (See Irving Kravis and coauthors, "Real GDP," pp. 215–242; and Kravis and coauthors, *UN International Comparison Project: Phase II; International Comparisons of Real Product and Purchasing Power* (Baltimore, Johns Hopkins University Press for the World Bank, 1978.) As a rough order of magnitude, the Kravis studies point to an increase in the percentage for the developing countries from 22 to 35. The definition of "developing countries" for this purpose is the World Bank's classification together with the "Asian centrally planned" countries—notably including the People's Republic of China.

The major contrasting group is the market system "industrial countries," sometimes defined as all twenty-four members of the Organisation for Economic Co-operation and Development (OECD) and sometimes excluding Mediterranean Europe (Portugal, Spain, Yugoslavia, Greece, and Turkey). South Africa is another borderline case. When the communist system countries are not separately recognized (as they generally are in United Nations data under the heading "centrally planned economies"), the Soviet Union and Eastern Europe are treated as "industrial" and China and the other Asian communist countries as "developing."

The World Bank in its *World Development Report, 1979*, tabulates as "developing countries" 95 countries with populations over one million and 27 smaller countries; of these 122, five (Saudi Arabia, Libya, Kuwait, Qatar, and the United Arab Emirates) are specifically identified as "capital surplus oil exporters." There is a separately classified group of 12 "centrally planned economies," of which the USSR and 6 in Eastern Europe resemble the western industrial countries in income levels and economic structures (Rumania being a borderline case), while those in Asia (China, North Korea, and Mongolia), along with Albania, are put in the general category of "developing countries." In Europe, Turkey, Portugal, Yugoslavia, Greece, and Spain are classified by the World Bank as "developing" rather than "industrialized," although the income levels and degrees of industrialization are not far from those of Ireland and Italy, the least affluent members of the European Community. Taiwan—formerly known as the "Republic of China"—is also treated as a separate "developing country" in the World Bank statistics. Including the capital surplus oil exporters and the poorer centrally planned economies, that classification results in a total of 127 developing countries and 27 industrial countries.

2

The Traditional Sector

One-quarter or more of the energy consumed in developing countries is in the form of traditional fuels, and perhaps half of the world's population—2 billion people—rely primarily on traditional fuels for their direct energy needs. This chapter describes these fuels, how they are used, and the problems encountered in their use or overuse. These include current local shortages which may become more acute in the future, and environmental hazards within local areas, some extending to global proportions.

Traditional Energy Use

Traditional Fuels

In the industrial countries and in the modern sectors of the developing countries, most of the energy consumed is in the form of commercial or conventional energy—petroleum, natural gas, coal, and electricity generated from fossil fuels, water power, or nuclear energy. Most people in developing countries, however, meet their household and production energy needs through noncommercial or traditional fuels and animate energy sources. The fuels are often gathered and used directly by family members without entering money markets. They include: (1) wood fuels—firewood and charcoal; (2) residue fuels—dung from cattle or other animals; and (3) crop "wastes" such as wheat straw and sugarcane bagasse. Wood is the primary fuel used in rural areas, with charcoal being more popular in urban areas because of its convenience and transportability. Crop residues and dung are usually resorted to only where wood fuels are unavailable or too costly.

45

A major area where commercial and traditional fuels interconnect and which may lead to additional problems is the progressive commercialization of traditional fuels. Our treatment of the traditional and modern sectors separately obscures the fact that there is no hard and fast distinction between them. The energy supply system should rather be thought of as a spectrum—ranging from the purely traditional fuels, which can only be gathered and consumed locally, through wood fuels that can be consumed locally but may also enter into organized commerce and be consumed many miles away, to the purely modern fuels, which are totally commercially organized and distributed to far distant markets. The mix of commercial and traditional fuels varies significantly according to relative prices and seasonal availabilities.

There may be considerable competition between wood fuels consumed locally and those delivered to urban areas. Indeed, there is evidence of a strong attachment to charcoal as a cooking fuel even when other more efficient fuels are available. This connection between the urban and rural use of similar fuels may be of particular importance now, as prices of modern fuels have increased, leading toward increased demand for wood fuels and charcoal in urban areas and intensifying supply problems in rural areas.

Since most traditional fuels are not traded in commerce, estimates of their consumption are necessarily very rough.[1] Nonetheless, estimates in the aggregate and by countries have been made by the United Nations Food and Agriculture Organization (FAO) and by the World

---

[1] Total potential supplies are not an accurate measure of consumption, since not all supplies are available for use (for example, forests may be long distances from population centers, or livestock- and land-produced residues may be owned or controlled by an upper class). Even if all supplies were available, other unknowns that must be determined for residues are the numbers of livestock and crop yields, the amounts of dung and crop residues produced per animal (which vary by type, size, and nutrition of livestock) and per ton of agricultural crop, the amounts actually collected, and the energy content of these residues, all of which may vary by a factor of two or more.

The consumption of wood fuels poses similar problems of measurement, since wood production has typically been measured for purposes of commercial forestry in terms of cubic meters of roundwood; yet all biomass may be burned as fuelwood or converted into charcoal, including branches, twigs, and leaves of trees and shrubs and other small plants. The species of tree, moisture content, and proportion of bark, leaves, and twigs also influence the heating value of wood.

The most reliable way of measuring traditional fuel consumption would be to survey actual use, although several of the measurement difficulties cited above would still be problems: for example, headloads and bundles of wood or crop residues are not readily translated into energy values. Few such surveys have been made, however, and those that have illustrate the difficulties of measuring traditional fuel consumption. A survey

Bank staff.[2] For 1978, the FAO estimated wood and charcoal consumption in all developing countries at 1,068 million cubic meters, or 262 million tons of oil equivalent.[3] The slightly higher World Bank estimates, together with some 400 million tons of animal dung (28 million tons of oil equivalent [mtoe]) and an unknown but perhaps similar quantity of crop residues, are the basis of the numbers shown in figures 0-1 and 0-2, amounting to 159 kilograms of oil equivalent per capita, or 28 percent of developing-country total energy consumption. Reliance is heavier on traditional fuels in poorer countries and in rural areas, among the urban poor relative to the urban non-poor, and geographically in Africa and Asia. Table 2-1 gives a general picture of the extent of national reliance on traditional fuels, while table 2-2 gives some estimates by geographical area and income class of the relative numbers of people using commercial energy, wood fuels, and dung and crop residues as their principal cooking fuels. Clearly, the majority of developing countries are heavily dependent on traditional fuels, and these are typically the poorer and more rural of the group.

It should be noted, however, that figure 0-2 overestimates the consumption of traditional energy in useful form, since the numbers are based on the heat content of conventional and traditional fuel sources. The efficiencies with which these fuels are used vary greatly and determine the amounts of useful energy that are actually produced and consumed. Efficiencies of use of traditional fuels—burned in open fires or primitive stoves—are typically very low, about 10 percent or less, so only a fraction of their available energy is actually consumed; while efficiencies of use of fossil fuels are often 50 percent or higher. More efficient stoves could increase the efficiency of use of traditional fuels but the price of these stoves is usually beyond the reach of many

---

in Upper Volta, for example, found a large seasonal variation in traditional fuel use, with crop residues being utilized conveniently from nearby fields after the harvest and wood being carried from further away at other times, showing the need for sustained surveying over time (Elizabeth Ernst, "Fuel Consumption Among Rural Families in Upper Volta, West Africa" [Ouagadougou, Upper Volta] Peace Corps, July 5, 1977).

[2] UN Food and Agriculture Organization, *Wood for Energy* (Rome, FAO, 1980) p. 17. The FAO estimates for wood fuels and charcoals are now also published as part of the United Nations statistical Series J, *World Energy Supplies*. For World Bank staff estimates, see Jyoti K. Parikh, *Energy and Development*, World Bank Public Utilities Report, No. PUN 43 (Washington, D.C., August 1978); and David Hughart, *Prospects for Traditional and Non-Conventional Energy Sources in Developing Countries*, Staff Working Paper No. 346 (Washington, D.C., World Bank, 1979).

[3] This assumes that 1 cubic meter of wood equals 232 kilograms of oil and 1 ton of wet dung and crop residues equals 70 kilograms of oil.

Table 2-1. Estimated National Reliance on Traditional Fuels, 1976

(each group arranged in ascending order of per capita GNP)

Modest reliance (less than half)	Medium reliance (approximately one-half to three-quarters)	Heavy reliance (three-quarters or more)
Pakistan (22)	Togo (67)	Benin (86)
Mauritius (2)	India (28)	Burundi (89)
Morocco (22)	Indonesia (62)	Cameroon (82)
Rhodesia (Zimbabwe) (36)	Sri Lanka (55)	Cape Verde (NA)
China (9)	Vietnam (55)	Central African Empire (91)
N. Korea (<1)	Gabon (44)	Chad (94)
S. Korea (8)	Liberia (53)	Ethiopia (93)
Philippines (<1)	Mauritania (63)	Gambia (73)
Ecuador (20)	Senegal (63)	Guinea (74)
Albania (24)	Zambia (45)	Guinea Bissau (87)
Algeria (4)	Thailand (34)	Kenya (74)
Tunisia (25)	Bolivia (45)	Lesotho (NA)
Iran (1)	Colombia (37)	Madagascar (80)
Lebanon (2)	El Salvador (53)	Malawi (82)
Argentina (3)	Guatemala (60)	Mali (97)
Chile (14)	Honduras (64)	Mozambique (74)
Cuba (5)	Malaysia (25)	Niger (87)
Dominican Republic (19)	Mongolia (25)	Rwanda (96)
Guadaloupe (19)	Brazil (38)	Sierra Leone (76)
Mexico (4)	Costa Rica (50)	Somalia (90)
Panama (29)	Nicaragua (47)	Sudan (81)
Peru (20)		Tanzania (94)
Uruguay (13)	(21 countries)	Uganda (91)
Fiji (2)		Upper Volta (94)
Cyprus (NA)		Zaire (76)
Malta (NA)		Afghanistan (76)
Portugal (3)		Bangladesh (63)
Romania (2)		Bhutan (NA)
Turkey (18)		Burma (85)
Yugoslavia (4)		Cambodia (93)
Libya (5)		Laos (87)
Hong Kong (NA)		Nepal (96)
Israel (NA)		Yemen (NA)
Singapore (NA)		Haiti (92)
Bahamas (NA)		Angola (74)
Venezuela (8)		Botswana (NA)
		Congo (80)
(36 countries)		Eq. Guinea (86)
		Ghana (74)
		Nigeria (82)
		Swaziland (NA)
		Paraguay (74)
		Papua New Guinea (66)
		(43 countries)

Notes for table 2-1 →

Table 2-2. Principal Cooking Fuels of World Populations, 1976

Populations	Percentage of population using			Total population (millions)
	Commercial	Wood fuels	Dung and crop wastes	
Africa South of Sahara	10	63	26	340
Urban nonpoor	83	17	0	30
Urban poor	0	100	0	20
Rural	3	66	31	290
India	10	48	43	610
Urban nonpoor	67	33	0	60
Urban poor	0	57	43	70
Rural	4	48	48	480
Rest of South Asia	12	46	41	205
Urban nonpoor	75	25	0	20
Urban poor	0	67	33	15
Rural	6	47	47	170
East Asia developing (Pacific)	36	42	23	265
Urban nonpoor	73	27	0	55
Urban poor	50	50	0	30
Rural	22	44	33	180
Asian centrally planned	22	51	27	855
Urban	73	27	0	205
Rural	6	58	35	650
Middle East & North Africa	53	18	3	200
Urban nonpoor	100	0	0	70
Urban poor	50	50	0	20
Rural	23	23	55	110
Latin America & Caribbean	71	26	3	325
Urban nonpoor	100	0	0	145
Urban poor	50	50	0	50
Rural	46	46	8	130
Total, developing countries	26	45	28	2,800
Developed countries (incl. Eastern Europe & USSR)	100	0	0	1,105
World	47	32	20	3,904

Note: These estimates are highly speculative and are only intended to give a general idea of orders of magnitude.

Source: Adapted from David Hughart, *Prospects for Traditional and Non-Conventional Energy Sources in Developing Countries*, Staff Working Paper No. 346 (Washington, D.C., World Bank, 1979) p. 4.

Notes for table 2-1

Notes: Country Reliance classified according to wood fuels plus estimated dung and crop wastes as a percentage of total energy consumption. Figures in parentheses are wood fuels alone as a percentage of total energy consumption in each country. Egypt, Iraq, Syria, Bahrain, Brunei, Kuwait, Oman, Qatar, Saudi Arabia, United Arab Emirates were not classified. NA = not available.

Source: William Knowland and Carol Ulinski, "Traditional Fuels: Present Data, Past Experience, and Possible Strategies," prepared for the Agency for International Development (Washington, D.C., AID, September 1979) pp. 9–10.

households. Cooking habits and social preferences may also make improved stoves unappealing to users.[4] After correction for relative efficiencies, the share of traditional fuels in terms of estimated useful energy delivered becomes less than 10 percent, rather than the 26 percent (based on heat content alone) shown in figure 0-2.[5]

On the other hand, the consumption of traditional energy may be underestimated because of probable underreporting of traditional fuels usage and especially because the estimates do not seek to include the considerable proportion of traditional energy that is provided by human and animal labor.

Animate Energy

A large part of traditional energy use is in the form of animate energy—animal and human labor.[6] This is particularly true in the agricultural sector, in which the majority of people in developing countries are employed. Measuring the amounts of animate energy used in tasks such as soil preparation, cultivation, harvesting, and transport is, however, at least as difficult as estimating traditional fuels consumption. One way is to measure the energy input of human beings and work animals on the basis of their food consumption. But not all food intake is used up in work. Most estimates conclude that no more than 5 percent of the total food intake of human beings and 10 percent of that of work animals is accounted for by useful work.[7]

Useful work energy is, therefore, a much smaller part of total energy consumption. In India, for example, where much of the research on measurement of animate energy has been carried out, such energy

[4] Elizabeth Cecelski, Joy Dunkerley, and William Ramsay, *Household Energy and the Poor in the Third World*, Research Paper R15 (Washington, D.C., Resources for the Future, 1979).

[5] This assumes average efficiencies of 10 percent for traditional fuels and 50 percent for conventional energy sources. Low efficiencies are not inherent in traditional fuels, but are related to the conversion devices used. An open fire has a very low efficiency, while a properly engineered wood stove may have an efficiency of 70 percent or higher.

[6] There is some difficulty with the inclusion of human energy inputs in the overall balance. It can be argued that one essential feature of the whole notion of development is the replacement of human by animal or mechanical power, so that human energy could be treated as a negative item rather than a positive one in composing an energy balance sheet relevant to development. We include human energy here nonetheless because we are interested in identifying tasks performed by human labor that might be more efficiently and comfortably performed using other sources.

[7] Ashok V. Desai, "Indian Energy Consumption: Composition and Trends," *Energy Policy* vol. 6, no. 3 (September 1978) pp. 221–222.

is estimated to account for 17 percent—3 for human beings and 14 for animals.[8] In agriculture, however, animate energy has been estimated to account in India for nearly 90 percent of total energy inputs.[9] Although less overwhelming, animate energy also appears significant in the agricultural sectors of many other countries.[10]

Rural Energy Use

Most energy in rural communities is locally produced from human and animal labor, wood fuels, and animal and crop residues, with "imported" (that is, from outside the localities) commercial fuels being used on a limited scale. Traditional fuels are usually gathered by family members, although wealthier families may purchase charcoal, dung cakes, or wood, and the poor may have to pay with services for the privilege of gathering firewood or residues on land that is privately owned. Much firewood is gathered not from forests, but from trees scattered along roads and fields, intercropped with agricultural crops, or in gardens and yards.

Estimates of consumption levels are highly uncertain, but results from a few limited surveys show per capita consumption of inanimate energy in rural areas as very low, between 70 and 250 kilograms (kg) of oil equivalent per capita in South Asia, 200 to 500 in Africa, and 350 to 800 in Latin America. These levels compare with an estimated minimum "adequate" consumption of 200 to 300 kilograms of oil equivalent per capita per annum for the provision of basic needs.[11] The share of household budgets taken by energy varies according to the availability of traditional fuels, but has been reported as low as 5 percent (similar to the industrial countries) in parts of Latin America and as high as 20 percent in India and Colombia, with the fraction of income spent on energy declining as incomes rise.

An estimate of typical energy use in rural India is given in table 2-3, showing the very minor role of commercial energy. In this case,

8 Ibid., p. 225.

9 Ibid., p. 227.

10 Roger Revelle, "Requirements for Energy in the Rural Areas of Developing Countries," in Norman Brown, ed., *Renewable Energy Resources and Rural Applications in the Developing World*, American Association for the Advancement of Science Symposium no. 6 (Boulder, Col., Westview Press, 1978).

11 These levels are based on the use of extremely inefficient energy conversion; they could be substantially reduced by using more energy-efficient methods of cooking (Cecelski and coauthors, "Household Energy," pp. 24–25, 32).

Table 2-3. Estimated Energy Use in Rural India
(kilograms oil equivalent per year per capita)

Source of energy	Agriculture	Domestic activities	Lighting	Pottery, brick making, metal work	Transportation and other uses	Total	Percentage
Local sources:							
Human labor	13.4	8.9	—	0.2	2.1	24.6	9.5
Bullock work	30.8	—	—	—	5.9	36.7	14.1
Firewood and charcoal	—	154.4	—	—	—	104.8	40.3
Cattle dung	—	—	—	17.0	—	42.3	16.3
Crop residues	—	—	—	—	—	24.34	9.4
Total	44.2	163.3	—	17.3	8.0	232.8	89.5
Commercial sources:							
Petroleum & natural gas:							
Fertilizer	8.0	—	—	—	—	8.0	3.1
Fuel	1.8	—	9.6	—	—	11.4	4.4
Soft coke	—	3.2	—	—	—	3.2	1.2
Electricity:							
Hydro	1.1	—	0.23	—	—	1.0	0.4
Thermal	2.7	—	1.1	—	—	3.9	1.5
Total, commercial	13.2	3.2	10.9	—	—	27.3	10.5
Total, local & commercial	57.4	166.5	10.9	17.3	8.0	260.0	100.0
Activities as a percentage of total energy consumption	22.0	64.0	4.0	7.0	3.0	100.0	100.0

Note: Dashes = not applicable.
Source: Adapted from Roger Revelle, "Energy Use in Rural India," *Science* vol. 192 (1976) p. 973.

local sources account for nearly 90 percent of all energy use, with domestic activities (mainly cooking) using 64 percent of total energy consumed. The main use of energy in rural areas is for cooking, with space heating and protection against predators as important secondary uses in some places. These uses are met almost entirely by traditional fuels. Human and animal labor are critical for gathering these fuels and for hauling water as well as for nearly all agricultural work. Dung from cattle, mainly used as fertilizer, is also subject to competing end-use as fuel. Small rural industries often use traditional fuels.

Commercial fuels, such as candles, kerosine, and electricity, are used by households for lighting and by wealthier families for ironing, radios, cooking, and so forth. Irrigation water, though usually supplied by gravity or animal power, is sometimes raised by electric or diesel pump sets.

A Village Energy Balance

The complexity of the rural food–fuel–livestock system is better understood by examining energy flows in one rural village, although this system of course may be very different in other villages. Dhanishwar, Bangladesh, is an example of an agricultural ecosystem in which little material output is wasted.[12] Figure 2-1 illustrates graphically the intricacy of the human, animal, traditional, and conventional fuel flows in this village and makes clear the "connectedness" among various sectors. A change in the number of livestock, the ratio of grain to straw (as through the introduction of high-yielding crop varieties), or the price of commercial fuels could upset this balanced subsistence energy economy. Crops produced by the village have several complementary functions and alternative outputs: *doinshah*, for example, a legume, edges individual paddy plots and keeps out water hyacinths during the monsoon, fixes nitrogen, and provides leaves for fodder and stems for fuel. Another crop, *amon* paddy, provides food, with its straw, husk, and leaves having alternative uses as fodder, fuel, fertilizer, and housing.

All labor in Dhanishwar is employed during the peak agricultural seasons; thus increased time spent in collecting fuel would probably reduce food production (and in fact, may already be a constraint).

[12] This description is based on John Briscoe, "The Political Economy of Energy Use in Rural Bangladesh," mimeo. (Cambridge, Mass., Harvard University Division of Applied Sciences, August 1979).

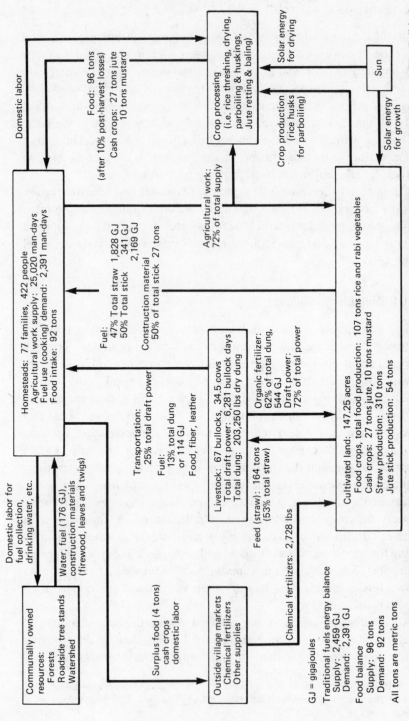

Figure 2-1. Aggregate Annual Energy-Related Resource Flows in the Village of Dhanishwar, Bangladesh, 1977. From John Briscoe, "The Political Economy of Energy Use in Rural Bangladesh," mimeo. (Cambridge, Mass., Harvard University, Division of Applied Sciences, 1979). Reprinted by permission of the author.

54

Besides providing one-half to two-thirds of the mechanical energy used in agriculture, as well as supplying transportation and dung for fertilizer or fuel, draft animals are straw consumers. An estimated 62 percent of all dung produced in the village is used as fertilizer and 13 percent is used as fuel, probably the maximum that is collectable. Other things being equal, dung burned rather than used as fertilizer may mean lower food production. Housing construction also competes with fuel for jute sticks and other crop residues, and rice straw is used as cattle feed and fuel.

As this study shows, most material outputs have competing uses, and any change in any part of the village ecosystem may affect energy availabilities. Population growth, weather variations, and many other factors can easily upset the delicate energy and subsistence economy balance.

Urban and Industrial Use

Small rural and urban industries, some large modern sector industries, and the urban poor are also important users of traditional fuels. In urban areas and for the urban sector, commercial fuels are easily available and more affordable, since incomes are higher. Traditional fuel use is not usually tightly tied to farming or other productive activities (though fuels may indeed be by-products, as in a sugar or sawmill), and monetary price and convenience become much more important considerations in the choice of fuels.

In urban households, commercial fuels are commonly used together with traditional fuels, both of which are sold in organized markets. Charcoal is generally preferred to wood in cities because of its convenience, compactness, and cleaner burning, and surveys in Asia and Africa have found per capita consumption of wood fuels (including charcoal) in towns to be higher than in the countryside, probably because of relatively higher incomes in urban areas.[13] In low-income urban areas, per capita demand for wood fuels can be quite high, caused especially by the greater use of charcoal, which usually requires a larger raw material input.[14] In 1972, for example, use

[13] Keith Openshaw, "Woodfuel, A Time for Reassessment," *Natural Resources Forum* vol. 3 (1978) p. 4.
[14] The higher efficiency with which charcoal is generally burned may offset conversion losses, depending on the charcoal-making process and on stove technology.

Table 2-4. Industrial Consumption of Traditional Fuels in Selected Countries, 1967–1977
(absolute figures in thousand metric tons oil equivalent (ttoe))

Countries	1967			1973			1976		
	ttoe	Percent total traditional fuels[a]	Percent total industrial energy[b]	ttoe	Percent total traditional fuels[a]	Percent total industrial energy[b]	ttoe	Percent total traditional fuels[a]	Percent total industrial energy[b]
Argentina	1,070	51	20	1,532	69	22	3,700	80	38
Brazil	2,825	12	28	4,459	19	23	4,166	15	17
Colombia	197	3	73	267	5	55	309	6	29
Egypt	120	84	17	189	87	21	190	87	14
India	778	3	3	1,316	5	4	1,661	5	4
Indonesia	203	1	58	289	1	46	455	2	40
Iran	151	29	9	215	45	2	215	31	2
Mexico	796	26	5	927	30	4	894	30	3
Thailand	153	52	20	456	84	19	869	93	28
Venezuela	131	8	3	277	14	4	300	14	4

Source: International Energy Agency/Organisation for Economic Co-operation and Development, *Workshop on Energy Data of Developing Countries*, vol. II *Basic Energy Statistics and Energy Balances of Developing Countries, 1967–1977* (Paris, OECD, 1979). Many of these figures must be treated with caution; a relatively large proportion of consumption of many fuels is often not allocated by sector.

[a] The percentage of all traditional fuel consumption that is consumed by the industrial sector.
[b] The percentage of all industrial sector energy consumption that is traditional fuels.

of wood in Bangkok was estimated at 3 million cubic meters, or almost 2 cubic meters per inhabitant.[15] As incomes rise, however, commercial fuels typically are substituted for wood fuels in urban areas.[16]

Industrial use of traditional fuels also appears quite extensive. Estimates of nonhousehold consumption of wood for energy in surveyed areas of Africa and Asia vary from 2 to 25 percent of total wood consumption.[17] As table 2-4 shows, in a number of countries industrial consumption of traditional fuels is not only large but rising, both absolutely and as a share of total traditional fuels use. (The share of traditional fuels in total industrial consumption, on the other hand, has generally decreased, reflecting the expansion of modern industry.) Some industries, sensitive to price changes, are likely to continue to rely on or even revert to use of traditional fuels in response to further price rises of fossil energy.

Wood and charcoal are used in brick and tile making, cement and metal industries, crop drying, bread baking, and fish curing. Tobacco curing appears to account for 17 percent of total energy consumption annually in Malawi, or 1 million cubic meters of fuelwood.[18] The Ugandan tea industry and railways in Thailand are also heavy users of wood fuels.[19] Other important industries using traditional fuels are some steel mills in Brazil, Argentina, and the Philippines, which use charcoal rather than coal, and sugar mills (and in some cases refineries), which are able to be self-sufficient in energy by using sugarcane wastes ("bagasse") to provide heat for boiling and evaporation and sometimes to produce electricity. Little information is available, however, on the efficiency of traditional energy use by industry.

---

[15] M. F. E. de Backer and K. Openshaw, *Present and Future Forest Policy Goals, A Timber Trends Study, 1970–2000,* Report No. TA 3156 (Rome, UN Food and Agriculture Organization, 1972).

[16] This distinction soon becomes difficult to make, however, since both traditional and conventional fuels are bought and sold in urban areas. The cost of stoves then probably becomes significant. In some areas, notably the Middle East, charcoal is preferred for its cooking qualities even when kerosine and other cooking fuels are available and affordable.

[17] J. E. M. Arnold, "Wood Energy and Rural Communities," paper presented at the Eighth World Forestry Congress, Jakarta, Indonesia, October 16–28, 1978, p. 5; Openshaw, "Woodfuel," p. 5.

[18] Arnold, "Wood Energy," p. 5.

[19] Openshaw, "Woodfuel," p. 5.

The "Other Energy Crisis": Pressure on Traditional Fuels

Local Signs of Stress

The "other energy crisis" in developing countries is the crisis not of oil supplies, but of traditional fuel availabilities and cost.[20] Shortages of traditional fuels are not new. But in the past, when relative "prices"—in terms of labor or other cost or difficulties in obtaining traditional fuel supplies—have increased, urban dwellers and industrial users have typically been able to shift to conventional, more efficient fuel sources, such as coal, water power, petroleum, or natural gas. Today, while the costs of traditional fuels are high in many areas, those of petroleum and other conventional sources are often even higher. One response has been to increase consumption of traditional fuels, further endangering the resource base and weakening prospects for economic development.

While biological energy resources are renewable, consumption in concentrated areas can erode the resource base that produces the renewable energy source. If more wood is taken from forests every year (through fuelwood gathering, charcoal making, timber and pulpwood production, or land clearing for agriculture) than the annual increment of forest growth, eventually the forest will disappear. This crisis can also cause spillovers in other sectors of the economy: if crop and animal residues are burned rather than returning nutrients and organic matter to soils, agricultural productivity can drop. Runoff and leaching of nutrients, as well as increased danger of flooding, are also serious side effects of overuse of forest resources.

Shifts to crop and animal residues. Wood is the preferred traditional fuel in developing countries, and in most cases, people only burn crop and animal residues when wood is unavailable or expensive. A half-day's walk from home—about 10 to 15 kilometers—appears to be about the limit for gathering fuelwood; where incomes do not permit purchasing "imported" fuelwood or charcoal, people must then resort to dung and crop residues as fuel. This practice can have severe effects on soil fertility and agricultural productivity. Crop wastes and animal dung return nutrients and organic matter to the soil and help retain

---

[20] The term "other energy crisis" is taken from Erik Eckholm, *The Other Energy Crisis, Firewood,* World Watch Paper No. 1 (Washington, D.C., World Watch Institute, 1975).

moisture, while crop residues also protect the soil from erosion caused by heavy tropical rains. As a maximum estimate, it has been calculated that one ton of cow dung burned means forgoing food grain production of 50 kilograms.[21]

Deforestation and desertification. Because firewood cannot usually be economically transported over long distances, large demands for woodfuels by urban dwellers and industrial users (including charcoal making) can quickly place stress on forests in local regions. In drought-prone regions such as the African Sahel, even relatively small and scattered populations can outpace the ability of fragile ecosystems to renew themselves sufficiently to sustain fuel production, and cities in these areas are particularly prone to creating desert-like conditions. While clearing for agricultural use and overgrazing are probably more important causes of deforestation and desertification, fuelwood collection has been a contributing cause in North Africa, the Indian subcontinent, and the Andean and Caribbean regions of Latin America.

Globally, 97 million hectares or 2 percent of forests were added between 1965 and 1975 (see table 2-5). One author has estimated an annual global wood increment of 6.6 million cubic meters, contrasted with estimated consumption of only 2.8 to 4.8 million cubic meters.[22] But much of this increment in wood and the increase in forest lands is in temperate forests or in areas that are otherwise inaccessible to developing-country populations. Tropical forests are under much greater stress, being lost at a rate of perhaps 16 million hectares annually.[23] These overall figures, moreover, do not reveal the situation in certain regions and countries where losses have been much more severe. Another difficulty is that much fuelwood is gathered from outside of forests, and these figures do not reflect gains or losses along roads or sides of fields, or of trees intercropped with agricultural crops. Furthermore, since trees take at least five to ten years to grow, waiting for actual decreases in forest stocks is likely to be too late to halt a deteriorating situation. And in many developing areas, the serious effects of overuse of traditional fuel resources are already evident.

[21] John S. Spears, "Wood as an Energy Source: The Situation in the Developing World," paper presented at the 103rd Annual Meeting of the American Forestry Association (Washington, D.C., World Bank, 1978) p. 5.

[22] Keith Openshaw, "Energy Requirements for Household Cooking in Africa with Existing and Improved Cooking Stoves," paper presented at Bio Energy Congress, Panel on Residential Fuels, Atlanta, Georgia, April 22, 1980.

[23] Openshaw, "Energy Requirements."

Table 2-5. World Forests, 1965 and 1975
(total figures in millions of hectares, per capita figures in hectares)

Countries	1965		1975		Percentage change 1965–75	
	Total (million hectares)	Per capita (hectares)	Total (million hectares)	Per capita (hectares)	(million ha)	(%)
All developing market countries	2,148	1.4	2,082	1.1	−66	−3
Africa	563	2.3	544	1.7	−19	−4
Near East	140	0.9	139	0.7	−1	−1
Far East	350	0.4	328	0.3	−22	−7
Latin America	1,055	4.3	1,029	3.2	−26	−3
Other developing	41	11.4	42	9.2	+1	+2
All centrally planned	1,095	1.0	1,150	0.9	+55	+5
Asian centrally planned	147	0.2	201	0.2	+54	+27
Eastern Europe & USSR	948	2.9	949	2.6	+1	0
Industrial market countries	816	1.2	925	1.2	+109	+2
World	4,059	1.2	4,156	1.0	+97	+2

Note: Figures may not add exactly because of rounding. More recent figures published by the FAO modify the 1975 figures, showing 647 million ha in Africa, 346 in the Far East, and 856 in Latin America, or a difference of 119 million ha in these regions rather than 66. However, the regions for the two series are not identical, and these revised figures are not available for other regions; they do not include a 1965 revision. Therefore, the earlier FAO figures are used here for both 1965 and 1975 for comparability. See J. P. Lanly and J. Clement, "Present and Future Forest and Plantation Areas in the Tropics" (Rome, UN Food and Agriculture Organization, 1979).

Source: UN Food and Agriculture Organization, *FAO Production Yearbook* (Rome, FAO, 1976 and 1975).

The destruction of forests and ground cover can have serious environmental consequences. Trees and other vegetation hold moisture and soil, reducing erosion and flooding, particularly on mountain slopes. Tree removal can eliminate topsoil in upland areas, silt over fertile valleys, destroy reservoirs by siltation, and reduce food production. Deforestation also releases the carbon dioxide stored in plants and reduces the capacity of the forests to reconvert carbon dioxide to oxygen, adding to a global problem that is not limited to developing countries.[24] If deforestation in some areas, however, is offset by reforestation in others—as was the case between 1965 and 1975—or if regrowth occurs in cutover areas, there should in theory be no effect on global atmospheric carbon dioxide. Regional climatic effects, however, might still be large.

The rural poor: a special hardship. The stresses discussed above strike the poor especially hard. It is the poor who are the largest users of traditional fuels and the poor who have the least access to their means of production (forest and agricultural land and livestock are usually owned or controlled by wealthier classes). For them, the only possible responses to difficulties in obtaining fuels are to reduce their energy consumption from already very low levels or to devote more time and labor to collection. Although conventional fuels might be cheaper in overall resource terms than traditional ones because of their higher efficiencies, and although better stoves could increase the amount of useful energy received from traditional fuels, the poor often do not have the money to buy an "expensive" ($3 to $10) stove, or will not do so when the alternative is an open fire.

When traditional fuels become increasingly scarce, they may begin to be bought and sold in markets and take a growing proportion of family incomes. Buying wood fuels in some cities in the wood-short Sahel in northwest Africa has been estimated to require one-fourth to one-third of the average laborer's income.[25] Distances of 100 to 500 kilometers and more have been reported for the transport of both fuelwood and charcoal to cities in a number of developing countries.

[24] Burning fossil fuels also contributes greatly to increased carbon dioxide which could result in worldwide climatic changes. The relative contribution of fossil fuel burning and deforestation is uncertain and is currently under investigation by a presidential commission and the National Academy of Sciences, as well as by individual scientists.

[25] William Knowland and Carol Ulinski, "Traditional Fuels: Present Data, Past Experience, and Possible Strategies," prepared for the Agency for International Development (Washington, D.C., AID, September 1979) p. 15.

For the poor, rising prices are often measured in terms of the additional labor and time required to collect fuel. In parts of upland Nepal and central Tanzania, 200 to 300 person-days of work per year are required to gather a family's fuelwood.[26] As mentioned earlier, in other cases landowners may exact more services from the landless in order to permit them to gather wood or crop residues on their property. The added burden and drudgery of this labor usually fall on women and children, the traditional gatherers of fuel.

Another response to rising fuel prices has been to reduce energy consumption and to lower nutritional standards. In areas of West Africa, the number of cooked meals per day has been reduced. And in some hill regions of Nepal and Haiti, fuel shortages even appear to have influenced the choice of crops produced in favor of those requiring less cooking.[27]

Where shortages of traditional fuels exist, the poor are forced to exert even greater pressure on the sustainability of renewable energy sources. People may cut fruit trees and other economically valuable species for wood, reducing food and income yields, especially where trees are intercropped (as when trees provide shade for coffee). Living trees, seedlings, and tree roots may be destroyed for fuelwood; leaves and grasses may be raked from hillsides leaving only bare earth; and crop and dung residues stripped from fields, reducing agricultural yields (and residues) and creating added pressures for expansion of agricultural land. In a vicious cycle, soil fertility and agricultural products are reduced, and fuel supplies become even more difficult to obtain.[28]

The Demand–Supply Balance: Future Trends

What are energy consumption levels likely to be in the future, and what part of these needs can be met with traditional energy sources?

Surveys in parts of rural Africa and Asia have shown wood fuels consumption varying from 1.3 to 2.1 cubic meters per capita annually. The same sources have estimated that wood fuels consumption may be increasing at 1.5 to 2 percent annually in the developing countries, based on current rates of per capita consumption and projected

---

[26] Spears, "Wood as an Energy Source," p. 11.

[27] Ibid.

[28] National Council of Applied Economic Research, *Domestic Fuels in India* (New York, Asia Publishing House, 1959) preface.

population growth.[29] As we have seen, current per capita levels of total inanimate energy consumption are very low in developing countries. Larger quantities of energy in some form will be required in the future, both to support expanding populations and to move beyond subsistence means of production. Mechanical energy for agricultural and industrial production, liquid fuels for transport, and higher quality forms of energy for household and commercial use will be needed to support economic development.

Several forces will be at work influencing levels of consumption of traditional fuels. On the one hand, together with industrialization, urbanization, and rising incomes, increasing energy consumption appears to have been tied closely to economic development. People in developing-country modern sectors probably use more energy inputs than their counterparts in the traditional sector—though this is not entirely clear, particularly if human and animal energy are included. What is certain is that they are getting more energy output—from more efficient forms of energy and more productive conversion devices. In any event, the move from traditional to conventional fossil fuels and electricity has in the past been associated with economic development. For this reason, the share of traditional fuels might be expected to decline as incomes rise.

However, rising prices of oil and other conventional forms of energy will undoubtedly slow this trend toward declining consumption of traditional fuels. Such prices may in fact divert consumption into new unconventional forms of energy which use traditional fuels more efficiently (some of which are discussed in chapter 9). Indeed, even in urban areas and in modern, formal sector industries, traditional fuels often compete favorably with conventional energy at higher prices. These factors would tend to increase the use of traditional fuels.

On the other hand, the use of traditional fuels may come to be limited by their rising price (in monetary or other terms) and decreasing availabilities if the productivity of the capital stock—forests and land—that creates them is reduced through overuse. Consumption of traditional fuels in that case will be limited eventually by the sustainability of their production and the competition of other uses for the basic resources of land and water. Potential supplies of traditional fuels overall are probably sufficient in many countries to meet future consumption needs, even allowing for new uses; but supplies may be

[29] Openshaw, "Woodfuel," p. 8.

available only at higher cost, and local shortages and environmental consequences are likely to become increasingly severe because of concentrated usage and socioeconomic constraints on availabilities. These prospects make it urgent to develop methods for expanding sustainable supplies by using traditional forms of energy more efficiently and at lower costs, and by improved tree-growing practices.

3

Commercial Fuels:
Uses and Problems

The division between the traditional and modern economic sectors[1] is paralleled to a large extent by a division between traditional and commercial fuels, with the former predominating in rural areas. In both areas, the demand for commercial fuels was stimulated by the drive for modernization. Furthermore in rural areas, as the use of traditional fuels became constrained by the kinds of problems described in chapter 2, the solution, in the era of cheap and freely available oil, appeared simple: switch consumption to commercial fuels as rapidly as possible. That solution is no longer a realistic option for many developing countries because of the major problems now posed by increasing reliance on commercial fuels.

These problems arise first from the strong demand for commercial fuels. Later chapters will discuss possibilities of moderating this demand in the future, but these possibilities must be viewed realistically within the context of strong historically increasing consumption levels of those fuels inherent in any form of economic development. Second, the most important and most rapidly growing commercial fuel is oil. The price of oil has risen sharply in past years and is expected to continue rising. For the countries we consider here, oil supplies are largely imported, adding strain to already extended foreign exchange budgets. Finally, higher prices of commercial fuels raise major problems

[1] Almost every developing country has at least a small modern sector, typically including its administrative capital, its ports or other major shipment points for international trade, and some industrial activity in mining, plantation agriculture, food processing, and manufacture of light consumer goods. The more industrialized developing countries have large urban industrial complexes, manufacturing both consumer and capital goods and providing an array of commercial services that make their modern sectors strikingly similar to those of the industrial countries.

65

of equity. Kerosine is the only form of lighting for those without electricity and is widely used in cooking by the poor.

Rapidly Rising Consumption

The demand for commercial fuels in developing countries has been rising very rapidly since the mid-1950s (see table 0-1). Rates of growth from the mid-1960s to 1973 were higher than in industrial countries and have remained high since 1973, in contrast to the industrial countries where the rate of increase in consumption has fallen sharply. In many developing countries, the rate of increase slowed after 1973 following the rise in energy prices that took place after that year, but consumption continued to rise (see table 3-1).

Of major importance, the share of oil in commercial fuel supplies rose sharply (see tables 3-2 and 3-3) to the detriment of coal, which has a significant though diminished role only in those developing countries

Table 3-1. Commercial Energy Consumption in Selected Developing Countries
(absolute figures in million metric tons oil equivalent)

Country	1960	1965	1970	1973	1978	Annual percentage increase 1960–73	Annual percentage increase 1973–78
Algeria	1.9	2.0	3.3	5.1	8.7	14.3	11.3
Brazil	15.9	20.2	29.8	42.5	62.4	7.9	8.0
Colombia	5.1	6.4	8.7	10.2	12.3	5.5	3.8
Egypt	5.2	6.2	6.2	7.1	12.5	2.4	12.0
India	41.9	57.0	66.3	61.1	77.3	2.9	4.8
Indonesia	8.2	8.1	9.6	11.9	27.8	2.9	18.5
Jamaica	0.5	1.0	1.7	2.7	2.6	13.9	−0.8
Kenya	0.8	0.8	1.0	1.2	1.4	3.2	3.1
Korea	4.3	8.4	17.3	22.2	34.2	13.5	9.0
Mexico	18.9	22.3	36.1	45.1	63.0	6.9	6.9
Nigeria	1.0	1.8	1.9	3.3	5.0	9.6	8.7
Philippines	2.7	4.7	7.5	8.6	10.7	9.3	4.5
Portugal	2.3	3.2	4.4	6.1	6.9	7.8	2.5
Thailand	1.2	2.7	6.1	7.8	10.0	15.5	5.1
Turkey	4.6	7.2	11.4	16.1	23.3	10.1	7.7
Venezuela	8.9	12.4	16.2	21.9	26.7	7.2	4.0

Note: For 1960, 1965, and 1970 figures were reported in coal equivalent; they were converted to oil equivalent at the rate of one ton of coal = 0.68 tons oil.

Source: United Nations, *World Energy Supplies, 1950–1974,* Series J, No. 19, and *World Energy Supplies, 1973–1978,* Series J, No. 22 (New York, 1976 and 1979).

Table 3-2. Share of Individual Fuels in Commercial Energy Consumption: World, Industrial, and Developing Areas
(percentage)

	1960				1978			
	Coal	Oil	Gas	Primary electricity	Coal	Oil	Gas	Primary electricity
World	52	32	14	2	32	45	20	3
All industrial incl. E. Europe	48	34	16	2	29	46	23	3
All industrial excl. E. Europe	40	38	19	3	23	52	22	4
All developing	71	25	4	1	46	43	8	3
Developing excl. Asian centrally planned economies	30	59	9	2	15	67	14	4

Source: United Nations, *World Energy Supplies, 1950–1974*, Series J, No. 19, and *World Energy Supplies, 1973–1978*, Series J, No. 22 (New York, 1976 and 1979).

Table 3-3. **Share of Individual Fuels in Commercial Energy Consumption in Selected Developing Countries**
(percentage)

Country	1960				1977			
	Coal	Oil	Gas	Primary electricity	Coal	Oil	Gas	Primary electricity
Algeria	19	81	—	—	2	52	46	1
Brazil	10	80	—	10	9	76	2	13
Colombia	34	54	7	4	21	56	16	7
Egypt	3	97	—	—	7	80	7	6
India	83	15	—	2	66	28	1	4
Indonesia	6	67	26	1	1	83	16	1
Jamaica	—	100	—	—	—	100	—	—
Kenya	5	90	—	5	3	91	—	6
Korea	83	16	—	1	41	58	—	—
Mexico	6	71	19	3	8	67	21	3
Nigeria	36	64	—	—	4	83	9	4
Philippines	5	93	—	2	2	94	—	4
Portugal	26	62	—	12	6	83	—	11
Thailand	—	100	—	—	1	95	—	4
Turkey	66	33	—	1	24	73	—	3
Venezuela	—	53	47	—	1	49	46	4

Note: Dashes = nil or negligible.
Source: United Nations, *World Energy Supplies, 1950–1974,* Series J, No. 19 and *World Energy Supplies, 1972–1976,* Series J, No. 22 (New York, 1976 and 1978).

which possess domestic resources—Colombia, India, Korea, and China. Oil is now the single most important fuel in developing countries, accounting in many cases for 80 percent of total commercial energy consumption—a higher proportion than in industrial countries. In those countries which produce natural gas, consumption has risen sharply and now accounts for a significant part of the total. Indeed, without the development of this gas and of hydroelectric facilities, the oil needs of developing countries would have been much higher than they are today.

The sharply rising commercial energy consumption, which occurred at a time of rapid economic growth, was expected. As incomes rise, the demand for commercial energy will rise rather faster for a variety of reasons. The early stages of development are usually defined by structural changes in the economy such as the commercialization of agriculture and a growing industrial sector. Most of the activities connected with these changes—including those in the transportation and domestic sectors—can only be carried out using commercial as opposed to traditional fuels, at least with current fuel-using techniques. Insofar as development is associated with accelerated urbanization, unavailability of traditional fuels may also lead to the use of commercial fuels for cooking and other household purposes. Furthermore, the typically rapid growth in thermal electricity generation contributes to the rapid expansion in commercial fuel consumption.

There was also some substitution of commercial for traditional fuels typified by the case of the replacement of vegetable oil and wood by kerosine for lighting, and wood and charcoal by kerosine for cooking.[2] But in most cases, the share of modern fuels in the total has increased because of the rapid growth of inherently modern fuel-using end-uses–industry, transport, and urban living.

Energy and the Urban Poor

The strong demand for commercial fuels that was manifested in the past and that has continued since the 1973–74 increase in oil prices raises crucial problems for developing countries. One of the most difficult is the impact of rising energy prices on special groups of so-

[2] Elizabeth Cecelski, Joy Dunkerley, William Ramsay, *Household Energy and the Poor in the Third World*. Research paper no. 15 (Washington, D.C., Resources for the Future, July 1979), p. 23.

Table 3-4. Monthly Per Capita Consumer Expenditures in India, 1970–71
(rupees, 1 rupee = 12¢)

Items	Monthly per capita expenditure class															
	0–8	8–11	11–13	13–15	15–18	18–21	21–24	24–28	28–34	34–43	43–55	55–75	75 & above	All classes
Food, total	4.79	7.72	9.48	11.23	13.21	15.45	17.57	20.11	23.44	28.04	33.68	41.71	65.02	34.04
Pan, tobacco, & intoxicants	0.23	0.29	0.36	0.42	0.50	0.59	0.70	0.79	0.92	1.21	1.55	1.93	3.43	1.58
Fuel & light	0.61	0.99	1.14	1.17	1.35	1.54	1.75	1.95	2.22	2.67	3.14	3.88	5.68	3.15
Clothing & footwear	—	0.08	0.12	0.11	0.16	0.23	0.35	0.45	0.77	1.22	2.33	3.59	10.77	3.10
Misc. goods & services, taxes, & durable goods	0.32	0.67	1.00	0.97	1.11	1.41	1.72	2.17	2.82	4.18	5.95	9.56	28.12	8.55
Rents	—	0.36	0.20	0.20	0.30	0.41	0.43	0.62	0.81	1.17	1.76	2.84	7.87	2.43
Total consumer expenditure	5.95	10.11	12.30	14.10	16.63	19.63	22.52	26.09	30.98	38.49	48.41	63.51	120.89	52.85
Fuel & light as percentage of total consumer expenditure	10.25	9.79	9.26	8.29	8.11	7.84	7.77	7.47	7.16	6.93	6.48	6.10	4.69	5.96

Note: Dashes = not applicable.
Source: India, Ministry of Planning, Department of Statistics, *National Sample Survey; 25th Round: July 1970–June 1971, Tables with Notes on Consumer Expenditure* (New Delhi, 1975) p. 143.

ciety—in particular the urban poor. In most cities they are obliged to use either commercial fuels or traditional fuels that have been commercialized. In either case, they must buy fuel rather than gather it.

Patterns of energy consumption by the urban poor vary from country to country depending on local situations, but in most areas certain petroleum products appear to be important for two purposes: kerosine as an illuminant in areas where there is no electricity, and gasoline and diesel fuel in public transportation for carrying people to work over long distances. These uses are reflected in subsidization of kerosine prices and some transportation fuels in many countries and in the reluctance of governments to let energy prices rise when this can possibly be avoided.

The importance of expenditures on fuel and light to the family budget of the poor in India is shown in table 3-4. In the lowest income groups, expenditure on fuel and light is second only to that on food and frequently is virtually the only cash expenditure apart from food. As incomes rise, so do expenditures on fuel and power, although typically more slowly than total expenditures, implying that fuel and power (like food) occupy a declining proportion of total expenditures. But for the lowest income groups, fuel and power account for a significant part of total expenditures. Consequently, increases in energy prices involve great potential hardship especially when accompanied by rising food costs.

Adjustments to Oil Price Rises

Rising oil prices—particularly abrupt increases—also lead to major adjustment problems for domestic and external economies. The world has recently experienced two very sharp increases in oil prices—in 1973–74 and 1979–80. Although the first increase was eroded by inflation between these two years, by early 1980 real oil prices were far above pre-1973 levels.[3] More important, oil prices are widely expected to continue rising in real terms. Such a development will cause continual adjustment problems for the oil-importing developing countries, threatening not only their own financial viability and living standards but also the stability of the international monetary system.

The effects of an increase in oil prices on oil-importing countries

[3] See Joy Dunkerley and John Jankowski, "The Real Price of Imported Oil," *Energy Journal* vol. 1, no. 3 (July 1980).

can be divided into short-run and long-run effects. In the first instance, as oil imports tend to be a significant part of total imports and as it is difficult to reduce oil consumption or to compress non-oil imports substantially in the short-run, the increase in oil prices leads to a sizable increase in total imports. In the absence of an equivalent increase in exports, trade balances deteriorate, and the resulting deficits have to be financed by borrowing or running down foreign exchange reserves.

Internally, rising oil prices lead to rises in prices of other goods. At the same time, the need for consumers to spend more for their energy services means that there is less money to buy other things, and in the absence of countervailing measures, aggregate demand falls. The rise in prices of imported oil, therefore, has simultaneous inflationary and deflationary repercussions on the domestic economy, always a confusing situation for policy makers. The combined inflationary and deflationary effects in turn aggravate balance-of-payments difficulties. For the developing countries, the prices of their imports from the industrial countries rise; but at the same time, the lower level of economic activity in industrial countries that was aggravated by the oil prices makes it difficult for developing countries to maintain export levels.

Following an increase in oil prices, therefore, the governments of oil-importing developing (and industrial) countries are faced with a variety of pressing problems—large trade gaps, inflation, and lower growth rates. In the longer term, the process of adjustment to the assumed permanently higher resource costs of imported oil involves a reduction in oil imports, either through increasing indigenous energy supplies or by conservation. These topics are the subject of subsequent chapters. Here we shall analyze, with reference to 1973–74, the short-run effects of oil price rises on balances of payments and domestic economies in order to assess the possible effects of subsequent oil price rises on economic peformance.

Effect on the Balance of Payments

For the oil-importing developing countries as a group, the rise in oil prices contributed substantially to the sharp rise in total imports. Both in 1974 and in the two subsequent years, however, oil imports accounted for only a part of the rise in total imports—37 percent in 1973–74 and 28 percent in 1974–77. Other factors were also clearly involved, some indirectly related to oil prices, such as the higher cost of imported manufactured goods, and some entirely independent.

Export revenues also rose sharply in 1974 and again, though less rapidly, in the two subsequent years. These increases in exports did not match the rise in imports, and the current deficit of the non-oil developing countries (balance on visible trade, including private but excluding official transfers) also rose (see table 3-5). From 1973 to 1974, the current deficit tripled. The deficit subsequently fell, but in 1977 was still three times the 1973 level (in nominal terms) when it was already considered high.

But as table 3-5 also shows, these large current deficits were more than covered by large increases in capital inflows. Indeed, foreign exchange reserves increased. Debt outstanding also rose; even so, by 1978, debt service, though rising, was still only about 13 percent of total exports of goods and services compared with 11 percent in the early seventies (table 3-6).

For the group of developing countries, therefore, the external adjustment to the 1973–74 oil price increase appears at first sight to have been surprisingly manageable, and certainly much easier than expected at the time. The unprecedented increases in the current account deficits of the non-oil developing countries as a group were smoothly financed by large capital inflows, the resulting high levels of debt remaining within operationally manageable levels. Does this experience mean that there are no further worries about external adjustment and that subsequent increases in oil prices can be accommodated without problems? There are a number of reasons for qualifying the success of the adjustment process of the late 1970s, reasons which have bearing on how the developing countries may adjust to similar shocks in the future.

To begin with, the analysis of oil-importing developing countries as a group can obscure important differences affecting individual countries. The group includes a wide range of countries with vastly different natural resource and energy bases, economic organization and structure, standards of living, and national aspirations. Given these differences, their reactions to a common shock such as the oil price rise and the nature of the adjustment to the common shock differ. Even though the group as a whole weathered the storm, individual countries encountered severe difficulties. In the future such difficulties could easily threaten the stability of the highly integrated international financial system.

Table 3-5. Balance of Payments: Non-oil Developing Countries, 1973–78
(billion U.S. dollars)

Year	Trade	Balance on services and private transfers	Current account	Capital account balance	Change in liabilities to foreign official agencies	Balance financed by transactions in reserve assets
1973	−6.6	−4.4	−10.9	18.5	0.1	7.7
1974	−22.2	−7.8	−29.9	31.4	1.4	2.8
1975	−28.6	−9.5	−38.0	37.0	1.7	0.7
1976	−15.4	−10.1	−25.5	33.4	3.4	11.3
1977	−12.2	−8.9	−21.2	32.4	0.4	11.6
1978	−21.6	−9.7	−31.3	44.1	−0.2	12.5

Source: International Monetary Fund, *Annual Reports, 1977, 1978, 1979* (Washington, D.C., 1978, 1979, 1980).

Table 3-6. Current Account Financing in Non-oil Developing Countries
(billion U.S. dollars)

Item	1970	1973	1974	1975	1976	1977	1978
Current account deficits		11.3	30.4	38.0	25.5	21.2	31.3
A. Financing not affecting debt		8.4	10.9	11.2	10.6	11.9	13.0
B. Net external borrowing[a]		10.6	22.4	27.5	26.2	21.0	30.8
from official sources		4.6	6.9	10.7	9.4	11.1	12.4
other[b]		6.0	15.5	16.9	16.8	9.9	18.4
C. Reduction of reserve asset[c]		-7.7	-2.9	-0.7	-11.3	-11.7	-12.5

	1970	1973	1974	1975	1976	1977	1978
For reference							
Debt service as a percentage of exports of goods & services	11.0	9.5	9.0	9.5	10.0	11.8	13.0

Source: International Monetary Fund, *Annual Reports, 1977, 1978, 1979.* (Washington, D.C., 1978, 1979, 1980).
[a] Public and publicly guaranteed borrowing only.
[b] Mainly from private banks.
[c] Minus signs = accumulation.

It it not possible to make here an individual examination of the external adjustment process for all the hundred or more oil-importing developing nations, but a more detailed analysis has been made for a selected number of countries.[4] These include very low income countries (India, Kenya) and somewhat higher income countries (Jamaica, Brazil, Portugal, and Turkey). These countries also cover a range of economic structures. Jamaica is a primary exporter; India, the Philippines, Portugal, and Turkey are balanced growth economies with a major agricultural sector; Argentina, Brazil, and South Korea have a greater degree of industrialization; and Kenya is an agricultural country.

For these individual selected countries, as well as for the oil-importing developing countries as a group, there is some indication that the short-run effects of the rise in oil prices on the external economy were handled without major disruption. These economies proved more resilient than originally feared. But it must be emphasized that our treatment of this topic was necessarily selective. Although representing a variety of income, economic structure, and energy consumption patterns, this selection does not cover those countries—including many in Africa south of the Sahara—which have the highest debt service ratios or whose prospects for further borrowing look most uncertain because of the fragility of their economies.[5] But on the surface, there was no clear connection between economic structure and success in adjusting to the first sharp increase in oil prices.

Future efforts to finance external deficits caused by rising oil prices may, however, be compromised by other factors. External borrowing has risen sharply—tripling in nominal terms between 1973 and 1978 (see table 3-6) with the shares of official and private borrowing in this

[4] Joy Dunkerley with Andrew Steinfeld, "Adjustment to Higher Oil Prices in Oil Importing Developing Countries," *Journal of Energy and Development* vol. 5, no. 2 (Spring 1980).

[5] Ibid., pp. 7–10. Moreover, such problems are not confined to small, poor countries. Some of the richer countries have experienced difficulties with their external deficits. A striking feature of the global balance of payments that often receives insufficient attention is the very high deficits recorded by the countries called in International Monetary Fund (IMF) nomenclature "more developed primary producing areas." These include Australia, New Zealand, South Africa, and within Europe, Finland, Greece, Iceland, Ireland, Malta, Portugal, Romania, Spain, Turkey, and Yugoslavia. Their combined deficit was equal to the combined deficit of the non-oil developing countries. Difficulties in financing external deficits in these countries could, by leading to a loss of confidence in the international credit structure, seriously affect the financial problems of the oil-importing developing countries. Defaults or a series of requests for debt rescheduling, even from small countries, could jeopardize the ability of other countries to borrow, no matter how prudent their own history of debt management.

total fluctuating widely from year to year. Although outstanding debt and debt service as a percentage of total exports in 1978 were not substantially above 1970 levels, a recent sharp spurt may presage further rapid increases. Given the erratic trends in the ratio since 1970, it is difficult to interpret this recent development, but in view of the recent increases in oil prices, debt service is expected to rise more rapidly in the future.

The debt data reported in table 3-6 apply only to guaranteed debt, that is, debt owed to foreigners that is guaranteed by public entities. This does not include private nonguaranteed debt, which is of major importance to some countries. The private sector in Brazil, for example, has obtained large amounts of external financing without public guarantee. If this were included in the data on debt servicing, the ratio would be even higher than the already very high 28 percent of exports in 1977.[6]

This rapid increase in borrowing, both guaranteed and nonguaranteed, particularly from the private banks, has raised several concerns.[7] One is, how long can it last? It is not surprising that developing countries have been borrowing heavily in private financial markets, because interest rates in that market have barely kept pace with inflation rates. The other side of the coin, however, is that lenders may not be willing to lend indefinitely at such low rates, especially if risks are felt to have increased. A further question is how far the private lenders may be overexposed, or are felt to be overexposed, in rapidly growing markets with which they are not very familiar. As little is known about the portfolios of the major private lending institutions, it is difficult to give a realistic assessment.

For some countries, moreover, the adjustment process in the 1970s may have been assisted by exceptionally favorable, but temporary, features of the world economy or by actions which cannot be sustained indefinitely. For example, some of the most successful adjustments came about through holding down import levels. In some cases, such as India, this was made possible through exceptionally good harvests. In others, it may have been achieved by holding down the imports needed for economic development, an action that will affect growth rates adversely in the future.

[6] Dunkerley with Steinfeld, "Adjustment to Higher Oil Prices," p. 9.

[7] See Robert Solomon, "A Perspective on the Debt of Developing Countries," *Brookings Papers on Economic Activity* vol. 2 (1977) for extended discussion of these points.

Furthermore, adjustment was facilitated through a rapid rate of increase in the volume of exports. But exports of the non-oil developing countries are highly dependent on market conditions in industrialized countries as evidenced by the fall in export volumes in 1975 during the recession in industrial countries. Some estimates indicate that the developing countries suffered more from the effects of oil-induced inflation and recession in the industrial countries than from the direct effects of the oil price increase.[8]

While this brief survey has dwelt mainly on the financial aspects of the question, it is important to recognize that political issues are also involved. High levels of debt imply diminished latitude in dealing with economic policy, which in turn can lead to major domestic political problems. It should equally be recognized, however, that the inability to borrow could also have major political repercussions. As seen in the section below, high levels of borrowing permitted developing countries to maintain relatively high growth rates. Without such growth the impact of higher oil prices on the poorer parts of the population would probably have been even more serious.

Looking to the 1980s, and the problems of adjustment to the doubling of oil prices in 1979–80 and further increases of unknown magnitude that may lie ahead, there are conflicting views as to whether it is realistic to expect a repetition of the relatively smooth accommodations of the seventies. The more optimistic school of thought argues that the scale of needed adjustment is relatively smaller than in 1974 and that the petrodollar surpluses are bound to find some channel to the deficit countries which once again will finance their continuing imports of oil and other requisites for development. Those holding the more pessimistic view note four differences from the past decade. (1) Starting from the plateau of high import levels, the OPEC surplus countries are unlikely to increase them further in a way offering developing countries new export markets and new emigrant worker jobs as a means of paying for part of their higher oil bills. An example here is Korea. The 1973–74 oil price rise presented no particular difficulty for this country

[8] See Edward R. Fried and Charles L. Schultze, editors of *Higher Oil Prices and the World Economy* (Washington, D.C., Brookings Institution, 1977). The authors estimate that "high oil prices reduced GNP by 2.5 percent in the U.S., 2.7 percent in Western Europe, and 4.2 percent in Japan. This slowdown in economic activity after a time lag works to reduce export prices and volumes in the developing countries, restricting in turn their capacity to import and to invest. At its peak, the depressing impact of these factors is likely to be somewhat larger than that exerted by higher oil import costs." (p. 34)

because Korean contractors with large-scale projects in the Middle East earned enough foreign exchange to pay for oil imports. Those contracts ran out in 1979–80, and the Korean economy is now likely to encounter difficulties paying for oil imports. (2) The supply–demand balance for oil is much tighter than in the mid-1970s, and the OPEC oil surplus countries seem intent on keeping it tight through unilateral production restrictions involving some form of tacit coordination, even if not formal production quotas; they are thus unlikely to permit the real value of oil to be eroded by general price inflation as it was in the 1970s. (3) The burden of indebtedness is therefore likely to increase sharply—both because the annual petrodollar surplus gap will remain large and because very high interest rates raise the cost of debt servicing even when the level of debt is unchanged. (4) Although it is axiomatic that petrodollar surpluses must be recycled in some form or other, there is no automatic mechanism to apportion those OPEC capital exports to the deficit countries most in need.

A balance-of-payments deficit cannot exist unless it is financed, either by drawing down reserves or by a combination of public and private borrowing from abroad. The potential crisis of the early 1980s is not only one of multiple defaults or near defaults, emergency debt rescheduling, and tightening of credit lines; it also consists of frustrated deficits—of desired imports of essential capital goods and raw materials required for economic development which must now be sacrificed or replaced by higher cost and less efficient substitutes. An overt financial crisis, involving a chain reaction of banking failures, can probably be warded off through cooperation among central banks and international financial institutions. Whether international cooperation can also be mustered to ward off a hidden development crisis is not so certain.

Effect on Growth and Inflation

Inflation and growth rates in developing countries were similarly affected by the increase in oil prices. The decade of the 1960s had been a period of relative price stability, at least by current standards, although there was considerable concern about inflation at the time. The GDP deflator for industrial countries rose at an annual rate of 4 percent from 1962 to 1972. In the developing world, inflation rates between 1967 and 1972 were substantially higher, about 10 percent (see table 3-7).

In the early 1970s, however, inflation rates had quickened and by

Table 3-7. Annual Average Inflation Rate, 1962–78
(percentage increase from previous year)

Countries	1962–72[a]	1973	1974	1975	1976	1977	1978
Industrial countries	4.1	7.3	11.9	11.0	7.1	7.1	7.0
Non-oil developing countries	10.1[b]	22.1	33.0	32.9	29.9	29.7	24.6

Note: Computed at compound annual rates of change.
Source: International Monetary Fund. *Annual Report 1979* (Washington, D.C., 1980) pp. 3, 11.
[a] Average
[b] 1967–72

Table 3-8. Annual Average Change in Output, 1962–78
(percentage change from previous year)

Countries	1962–72[a]	1973	1974	1975	1976	1977	1978
Industrial countries	4.6	6.1	0.2	−0.9	5.4	4.0	4.0
Non-oil developing countries	6.1[b]	7.3	5.3	4.1	5.0	5.1	5.2

Note: Computed at compound annual rates of change.
Source: International Monetary Fund. *Annual Report 1979* (Washington. D.C., 1980) pp. 3, 11.
[a] Average
[b] 1967–72

1973 were running in both groups of countries at double the rate of the previous decade. As early as the end of 1973, many countries considered inflation to be their major economic problem. In this context, the sharp rise in prices of oil appeared particularly alarming. Consumer prices rose sharply in 1974, often to a level 50 percent higher than 1973; while inflation rates subsequently diminished, prices continued to rise at rates well in excess of the pre-1973 period. The increase in oil prices provided a major contribution to this acceleration in inflation rates.

The increase in oil prices also contributed to a lowering in growth rates in the oil-importing developing countries (see table 3-8). Up to 1973, the oil-importing countries had achieved historically high rates of growth, culminating in a particularly rapid increase of 7 percent in that year. From 1974 to 1978 growth rates declined to 5 percent a year. These lower growth rates were still, however, much higher than those achieved in the industrial countries. Indeed the lower growth rates of the industrial countries contributed further to the difficulties of the oil-importing developing countries by restricting growth in their export markets.

It is at first sight surprising that the decline in growth rates in the oil-importing developing countries was not greater. The single most important factor in this better-than-anticipated performance appears to have been the high level of foreign borrowing which compensated for the deflationary effects of increased oil bills. In the future, as we have seen above, it may be more difficult to secure foreign loans on the scale required to ensure satisfactory growth rates. And, even though the oil-importing countries have so far managed to keep growth rates up, this achievement should not obscure the fact that *any* reduction in growth rates in poor countries with rapidly growing populations reduces the already too modest improvement in the living standards of the individual citizen. Higher oil prices, no matter how skillfully accommodated, add to the already heavy burden of achieving the necessary increases in economic output.

4

Energy, Economic Growth, and Development Strategy

Previous chapters have analyzed the typical consumption patterns of energy in developing and industrial countries, identifying both similarities and contrasts. One set of contrasts arises from the importance of traditional fuels and the problems involved either in their continued use or in their replacement by commercial fuels when cheap oil is no longer available. Another set of contrasts arises from the differing phases of development.

It was shown in chapter 1 that not only per capita levels of energy consumption but also energy intensities—the amount consumed per unit of economic output (sometimes expressed as the E/GDP ratio)— are generally lower in developing than in industrial countries. The changes in economic structure historically associated with development include enhanced agricultural productivity, growth of manufacturing, urbanization, and increased transportation of goods and services—all relatively energy-intensive activities. At a fairly high level of income per capita, however, these trends give way to an increase in services, and also to more sophisticated manufacturing in which energy and other material inputs account for a declining share of the final value of the product.[1] Especially since the abrupt rise in oil prices in 1973–74, the E/GDP ratios in the industrial countries have fallen substantially, and most projections and energy policy discussions anticipate their further decline.

For countries in the earlier phases of modernization, on the other

[1] See Hollis Chenery and Moises Syrquin, *Patterns of Development 1950–1970* (London, Oxford University Press for the World Bank, 1975); Daniel Bell, *The Coming of Post-Industrial Society* (New York, Basic Books, 1973); and Lincoln Gordon, *Growth Policies and the International Order* (New York, McGraw-Hill for the 1980s Project/ Council on Foreign Relations, 1979).

hand, it is generally expected that energy use—especially in its commercial forms in the modern sector—will intensify, notwithstanding higher relative costs. The experience since 1974 bears out that expectation. It is not implied that all developing countries must come to duplicate in all respects the production patterns and living styles of the presently industrialized countries. But for most of the poorer countries, their development aspirations clearly necessitate the expansion of industrial and agricultural output more rapidly than population growth, the development of modern skills and services, the building of transportation and communications networks, and the associated improvements in economic productivity.[2]

What is the relevance of energy supply and use to these changes? To what extent do costs and availability of fuels constrain the rate of economic growth or force development into particular patterns? What are the relationships between energy policies and alternative development strategies? Are there particular aspects of development on which energy policies might exert a decisive influence? These are the issues briefly explored in general terms in this chapter. Their detailed exploration depends on the energy resource situation and specific development conditions of each country.

History and Theory

The theory of energy's role in economic growth has not attracted much attention until recently. For historians of social modernization and technological change, energy has always stood out as a crucial factor in development. Control of fire and domestication of animals were milestones in the evolution of human society, along with the introduction of settled agriculture. Replacement of arduous human labor by animal or mechanical energy seems to incarnate the very essence of development. Yet in the classical economic theories of growth, energy was not even mentioned as a discrete input to production, much less

[2] It is, of course, possible that development success will come to be redefined in terms other than improvements in per capita consumption of goods and services. As a measure of economic or social welfare, GNP per capita is recognized to be defective on several counts, including its neglect of income distribution and its exclusion of environmental quality and other collective goods not traded in markets, to say nothing of noneconomic goods such as dignity, liberty, or participation. Developing countries and communities will, in any event, display a variety of preferences concerning social organization, methods of development, and priorities.

a potential determinant of growth rates. The basic factors of production were identified as labor, capital, and land, with raw materials later treated as an extension of the concept of land.

The reason for this neglect is simple. In the world of ideas, energy was not yet a recognized concept in the time of Adam Smith, Malthus, or Ricardo, and scarcely even by the time of John Stuart Mill. In the practical world, energy was not a perceived constraint, in contrast with land in a crowded Europe mainly engaged in agriculture. Ricardo's *Principles of Political Economy* appeared in 1817, seven years before Sadi Carnot's *Reflections on the Motive Power of Heat,* while the second law of thermodynamics was not clearly formulated until 1851. The practical application of electricity to lighting and motive power came only in the 1870s and 1880s, long after the writings of Mill.

Even in this period, however, there was some concern about the adequacy of natural resources, including energy. Depletion of forest supplies of firewood had been observed in England in the late eighteenth century. The prospective exhaustion of coal was the centerpiece of a famous warning by Stanley Jevons in 1865. Later in that century the conservation movement developed in the United States, with primary emphasis on limiting renewable resources to levels of sustainable yield and avoiding wasteful methods of mineral exploitation which would make large fractions of the deposits permanently irrecoverable. Only in the present century, however, were there developed systematic formulations of resource economics, including theories of pricing and exploitation rates for depletable resources and intergenerational equities in their usage. In these writings, however, emphasis was on natural resources in general rather than on energy. Barnett and Morse, for example, in their classic work *Scarcity and Growth* did not single out energy for special attention.[3]

There were important historical (as distinct from theoretical) studies of energy usage and shifts in supply sources, with analysis of resource prospects and policies to improve the supply base. But with little policy guidance, a set of fairly smooth transitions was taking place in the industrial countries—from wood to coal to oil and natural gas— each substitute being more efficient and more readily produced and transported than its predecessor. Electrification was becoming the universal mode of lighting, the main source of industrial and residential

[3] For references to the literature, see Harold J. Barnett and Chandler Morse, *Scarcity and Growth* (Baltimore, Johns Hopkins University Press for Resources for the Future, 1963) pp. 44–48.

motive power, and a substantial factor in intraurban transportation and some railway systems. During the first half of this century, the cost of energy materials in the United States, already small, fell further, from 2.6 to 2.0 percent of GNP, while the efficiency of energy conversion and use was drastically improved.[4]

Perhaps if energy costs had accounted for a larger share of the total, economists might have recognized earlier some of the characteristics that differentiate energy from other raw materials. There are three of special importance: (a) pervasiveness, a quality that implies some use of energy in all transformations of goods and the provision of all types of services; (b) the low ease of substitution, especially in the short term, when one fuel may replace another but total energy use is much harder to change; and (c) the impossibility of recycling energy—a direct corollary of the second law of thermodynamics. It is only in very recent years, mainly since the oil price crisis of 1973, that energy has been analyzed as a factor of production alongside of capital and labor. Attempts have been made to measure the extent of its complementarity or substitutability with the other productive factors, in order to assess how energy prices and availabilities influence overall rates of economic growth.

The results of several recent studies, based largely on U.S. data, indicate that rises in energy prices should not have a strong effect on long-run growth in industrial countries because the energy sector is a small component of the total economy and there appears to be substantial scope for substituting other inputs as energy prices rise.[5] Macroeconomic analysis bears out these comforting conclusions. Noting that the cost of primary energy in the United States amounted to 2.4 percent of GNP in 1972 and 5.2 percent in 1975, a common scenario assumes another doubling to about 10 percent of GNP. It is argued that the outside limit of the macroeconomic effect would be a 5 percent reduction in GNP; spread over twelve years, that would reduce an otherwise assumed growth rate of 3.5 percent a year to 3.1 percent.[6] Considering the possibilities of substitution of capital and

[4] Calculated from data in U.S. Department of Commerce, Bureau of the Census, *Raw Materials in the United States Economy 1900–1969* (Washington, D.C., 1972) pp. 11–13; and *Historical Statistics of the United States: Colonial Times to 1957*, also by the U.S. Department of Commerce (Washington, D.C., 1960).

[5] National Academy of Sciences, National Research Council Committee on Nuclear and Alternative Energy Systems (CONAES), *Energy Modeling for an Uncertain Future* Supporting Paper 2 (Washington, D.C., 1978) pp. 44–47 and 106–115.

[6] See Hans S. Landsberg, and coauthors, *Energy: The Next Twenty Years*, a study

labor for energy, the actual effect should be substantially less. Similar conclusions are sometimes stated in terms of the small loss in real income expected to result from reductions in energy growth rates, or even zero energy growth. It should be emphasized that such calculations are based on gradual price rises spread over sufficient time to permit changes in technologies and alterations in the capital stock; they exclude the dislocations and economic losses occasioned by abrupt changes such as those of 1973 and 1979.

As a theoretical matter, a similar change in energy costs in developing countries might have a more severe effect on economic growth, especially if the energy supplies are mostly imported and if the imports have to be financed currently. For most developing countries, the availability of capital and of foreign exchange for essential imports is the tightest constraint on their short-term growth rates. Moreover, any given reduction in overall growth rates falls more heavily on developing countries because higher population growth absorbs a larger fraction of the total than in industrial countries, leaving less for improvements in per capita incomes and living standards.

Consider a developing country whose principal source of commercial energy is imported oil, and whose oil imports amount to 4 percent of GDP (a figure near the 1976 experience for a representative group of countries). Suppose that the price of oil is doubled and that the extra cost is financed by expanding current exports. In effect, there has taken place a deterioration of the country's terms of trade, which now require 8 percent of GDP to pay for imported oil instead of 4 percent. To the extent that energy demand is reduced by higher prices—either through reduced ultimate consumption or through substitution of labor or capital for energy—there will be a smaller volume of imported oil and a smaller effort required to enlarge exports. Those substitution effects would reduce the impact on GNP to less than 4 percent. But other factors may worsen the effect on growth rates.[7]

If the levels of consumption are so low that they cannot readily be further compressed, the extra exports will be mainly at the expense of

sponsored by the Ford Foundation and administered by Resources for the Future (Cambridge, Mass., Ballinger, 1979) p. 20; and Sam H. Schurr, Joel Darmstadter, Harry Perry, William Ramsay, and Milton Russell, *Energy in America's Future: The Choices Before Us* (Baltimore, Johns Hopkins University Press for Resources for the Future, 1979) pp. 90–100.

[7] See Charles Blitzer, "Energy and Development," mimeo. (World Bank, Economic Analysis and Projections Department, December 1978) for further development of this argument.

savings and investment. Investment rates for developing countries are typically in the range of 15 to 20 percent of GNP including depreciation, and 12 to 16 percent net of depreciation. If three-fourths of the extra effort to pay for oil imports were to come out of investment, the net investment rate would be reduced by 20 to 25 percent (3 percentage points out of 12 to 16), leading to reductions in growth rates of a magnitude which could eliminate improvements in per capita income levels. In many cases, moreover, the capacity to import has been considered an even tighter constraint on growth than the overall level of investment, since continuing growth may require the importation of capital goods, raw materials, or components not available in the country concerned and not readily replaceable through substitution. If exports cannot be readily expanded and the only means to finance additional costs of oil imports is compression of non-oil imports, the adverse effect on growth rates may be very severe indeed.

The Experience of Developing Countries

How well does the recent experience of developing countries match these theoretical considerations? As a group, it was shown in chapter 3 that their economic growth suffered relatively little from the oil price increases of 1973–74. But a major factor in that outcome was the financing of continued high levels of oil imports through an enormous increase in international indebtedness, in effect deferring the impact for a few years. In other respects, the experience was more in line with expectations.

Trends in Energy Intensities

In the period before 1973, the cost of energy was a somewhat lower fraction of GDP in the developing than in the industrial countries. That was a natural corollary of lower energy intensities (or E/GDP ratios), since international energy prices tend to be similar for all countries. After the sharp price increases of 1973–74, however, energy costs in developing countries rose on account of both price and quantity, with the quantity increases concentrated on the side of commercial fuels. As shown in table 4-1, the increase in commercial energy consumption for the developing world outstripped the increase in economic growth, reflecting a continued rise in energy intensities. These increases

Table 4-1. World Economic Growth and Commercial Energy Consumption, 1970–78
(percent per year)

	Economic growth rates		Growth in commercial energy consumption		Change in energy intensity	
	1970–73	1973–78	1970–73	1973–78	1970–73	1973–78
World	5.6	3.4	4.8	2.5	−0.8	−0.9
All industrial countries[a]	5.4	2.9	4.1	1.5	−1.3	−1.4
All developing countries[b]	6.7	5.3	8.0	7.3	1.3	2.0

Sources: Derived from data in World Bank, World Economic and Social Indicators (Washington, D.C., 1979); United Nations, World Energy Supplies 1973–1978, Series J, No. 22 (New York, 1979).
[a] Western and Eastern Europe (including USSR), United States, Canada, Japan, Australia, and New Zealand.
[b] All others, including OPEC members, China, and other Asian centrally planned economies.

contrast with the industrial countries, where energy intensities were declining both immediately before and after 1973. It is too early to assess the effects of the 1979–80 oil price increase on energy intensities.

Data for a large group of disparate countries, however, may obscure important differences in experience among individual members of the group. Accordingly, commercial energy intensities for selected countries (including poor and less poor countries covering a wide range of economic development) are given in table 4-2. Inputs of commercial energy into the economies of most of these countries rose between 1960 and 1973, and for a lesser, but still substantial number, they continued to rise between 1973 and 1978.[8]

[8] The relationship between income (GDP) and commercial energy consumption in developing countries can be estimated more formally by calculating the income elasticities of demand for energy through regression analysis. The income elasticity gives the percentage increase in per capita energy consumption associated with a given percentage increase in per capita GDP. These calculations confirm the increasing commercial energy intensity of developing countries. Thus one study on "Energy Demand in Developing Countries," by B. J. Choe in Joy Dunkerley, ed., International Energy Strategies: Proceedings of the 1979 IAEE/RFF Conference (Cambridge, Mass., Oelgeschlager, Gunn, & Hain, 1980), covering three groups of developing countries (all developing countries, net oil-importing developing countries, and net oil-exporting developing countries) yields income elasticity of demand for commercial energy varying from 1.2 to 1.5; that is, commercial energy consumption (on past experience at least) has increased more rapidly than GDP, leading to a more energy-intensive economy. This result is also consistent with a cross-sectional analysis made at Resources for the Future of energy consumption and economic output (measured by purchasing power parity rates of exchange) among a group of twenty-six developing countries in the year 1973, which yields an income elasticity for commercial energy alone of about 1.3.

Table 4-2. Economic Growth and Commercial Energy Consumption in Selected
Countries, 1970–78
(percent per year)

	Economic growth rates		Growth in commercial energy consumption		Change in energy intensity	
	1970–73	1973–78	1970–73	1973–78	1970–73	1973–78
Algeria	5.3	5.5	14.4	11.3	9.1	5.8
Brazil	12.8	7.2	12.8	7.9	0.0	0.7
Colombia	7.2	5.6	5.4	3.9	−1.8	−1.7
Egypt	4.4	11.1	5.6	11.9	1.2	0.8
India	1.8	4.4	4.3	4.8	2.5	0.4
Indonesia	8.7	6.7	13.4	18.4	4.7	11.7
Jamaica	4.6	−2.5	4.5	−0.6	−0.1	1.9
Kenya	5.5	4.8	8.8	3.7	3.3	−1.1
Korea	9.7	10.3	5.8	9.1	−3.9	−1.2
Mexico	6.0	3.5	7.3	6.9	1.3	3.4
Nigeria	7.0	7.0	20.1	8.5	12.9	1.5
Philippines	6.8	6.1	3.9	4.5	−2.9	−1.6
Portugal	8.9	1.6	15.5	2.3	6.6	0.7
Thailand	6.9	7.3	11.8	5.1	4.9	−1.8
Turkey	7.7	6.1	11.5	7.7	3.8	1.6
Venezuela	4.4	7.7	7.1	4.0	2.7	−3.7

Sources: Derived from data in World Bank, World Economic and Social Indicators (Washington,
D.C., 1979); United Nations, World Energy Supplies 1973–1978, Series J, No. 22 (New York, 1979).

At the same time that commercial energy consumption relative to
GDP was increasing, consumption of traditional fuels relative to GDP
was decreasing. Whether total energy intensity increased or decreased
depends on which trend dominates. In an analysis of ten developing
countries with data on traditional as well as commercial fuels, it was
found that total energy intensities increased in about one-half of the
countries. If an adjustment is made to take into account the higher
efficiency of commercial fuels, however, the tendency toward increasing energy inputs relative to GDP becomes more pronounced.

The relative vulnerability of any individual developing-country
economy to international oil price rises, therefore, depends on its stage
of development as well as its resource endowment and degree of
import dependence. An agricultural subsistence economy with low
total energy consumption per capita derived mainly from traditional
fuels will be comparatively little affected. The greatest vulnerability
comes precisely when development is most dynamic—the period of
rapid structural change, increases in energy intensities, and shifts from
traditional to commercial fuels.

Substitution Possibilities

The effect of higher energy prices on economic growth can be partly offset if it is technically and economically possible to substitute other factors of production, such as labor or capital, for the high-priced energy, or other and less costly forms of fuel for high-priced petroleum.

Some considerations suggest that these sorts of possibilities in developing countries are not as extensive as in industrial countries. The modern sectors of developing-country economies are likely to be less flexible, with a narrower spectrum of industrial activities and fewer broadly educated managers and technicians with facility in changing products and processes. As a result of the far lower living standards of the developing countries, some of the typical methods of energy conservation in the industrial countries are not open to them; there is, in effect, much less inadvertent "waste." Most of the developing countries are in mild or warm climates, with little or no energy devoted to space heating, which offers many opportunities for conservation in colder regions.

On the other hand, there are certain characteristics of developing countries—their smaller energy-using infrastructure and their more rapid rates of growth—which give them, potentially at least, greater flexibility in choice of capital stock and development strategies. But they may not find it easy to take full advantage of such choices because of shortages of capital and foreign exchange. On balance, it seems prudent to assume in practice fairly limited substitution possibilities, as least for the medium term.

It follows, therefore, that the search for alternative energy sources and for measures to improve the efficiency of energy use is of high importance for the developing countries. It also becomes of great interest to ascertain the energy implications of alternative development strategies. The prospect of continuing energy price rises and uncertainty over physical supply availability may cause planners in developing countries to reassess development plans with a view to moving toward a less energy-intensive development pattern.

Energy Policy and Development Strategy

The energy sector is an increasingly important component of the economy, and its management (or mismanagement) can be critical to development prospects. At the same time, it is important to avoid a disproportionate or obsessive concentration on the energy sector, and equally important to avoid the isolation of energy sector planning from broader development goals and strategies. Recognizing that the changing economics of energy may entail new constraints, especially through pressure on available capital supplies or the capacity to import, energy planning should nonetheless be subordinated to overall development planning. Its central purpose is to identify the most effective policies to bring about a least-cost development of the energy sector that will fit the general development strategy.

One key question is the extent to which alternative development strategies involve characteristic energy implications—that is, is one strategy more energy intensive than another, or does it require energy in particular forms? The range of strategies discussed in much of the contemporary literature includes the following: (1) import-substituting industrialization; (2) export-oriented industrialization; (3) balanced agriculture and industry, with emphasis on comprehensive rural development; (4) priority for "basic human needs"; and (5) radical alternatives such as "another development." When the politico-economic history of any specific country's development performance is studied closely, there never seems to be a complete fit with any of these theoretical paradigms. Such paradigms should be thought of as main directions of strategy, rather than pure alternatives. It must also be recognized that, in many cases, national leaders do not have a deliberate strategy or are unable to get the necessary implementing measures adopted; in those cases, operative strategy has to be inferred from the country's performance rather than from official goals.

Industrialization Strategies

Some type of industrialization is inherent in economic development; it is the most characteristic structural change in the historic process of economic modernization. The two named industrialization strategies, however, refer to fairly well-defined alternatives adopted by different groups of countries in the 1950s and 1960s, both having in common so strong an emphasis on industry that agriculture was often left to stagnate or even to decline. Industrialization is normally associated with urbanization, and both processes have obvious consequences for energy demand patterns, leading to a rapid rise in the use of commercial fuels.

As between import-substitution and export-orientation, however, there are no obvious general reasons for one or the other to be systematically more energy intensive or to require energy in particular forms. The answer for any specific country will depend on what imported products are being replaced by domestic production or what new exports are being developed. An analysis of fragmentary data on energy intensity changes in two sets of countries—one emphasizing import substitution in the 1950s but shifting toward export orientation after the mid-1960s, and the other adopting export orientation strategies at an earlier phase—fails to show any systematic relationship.[9] The dynamics of import substitution have a built-in tendency toward increasing energy intensity as the process develops, since the first stage normally consists of final assembly from imported components or semifabricated materials, in which the bulk of the needed energy expenditure has already been embodied. As the process of substitution moves back up the production chain (for example, to steel fabrication and even steel mills), it is likely to incorporate the more energy-intensive phases.

As to export-orientation, the initial product mix is presumably based on the comparative advantage of the exporting country, reflecting its natural resources, location, labor skills, and wage rates, among other factors. To the extent that relative energy costs are large enough to affect competitiveness in world markets, they should be automatically incorporated into the determination of comparative advantage. But the dynamics of export-oriented industrialization strategies tend to push toward further fabrication of domestic raw materials in order to capture for the home country a larger share of the value added in the various stages before final consumption. The extent of such forward fabrication may strongly affect the degree of energy intensity. In the case of copper ore or bauxite, for example, moving to the stage of electrolytic refining in the country of origin involves a large increase in energy intensity, while for petrochemicals the shift forward into plastics and manufactured products lowers the average intensity.

To the extent that export orientation provides a better trade balance than import substitution, it will also widen the country's energy supply options by facilitating energy imports. Japan and Germany have provided dramatic examples in the late 1970s of the possibility of maintaining trade surpluses in the face of very large volumes of oil imports. Brazil, India, and Korea have also been able to offset

[9] See appendix 4-A to this chapter, "Energy Demand Comparisons Between Import-Substituting and Export-Oriented Countries."

substantial shares of their higher oil bills by increasing their exports. The broader issue of energy self-sufficiency versus continuing imports is discussed later in this chapter.

Balanced Agriculture and Industry

In the mid-1960s, many development analysts and planners turned away from industrialization as their cardinal objective, on the grounds that it could not create enough employment to keep pace with rapidly growing populations and labor forces; that it was fostering excessive urbanization, with massive slums and grossly inadequate urban services; and that the concomitant neglect of agriculture was leading to dangerous and costly overdependence on food imports. The preferred development strategy advocated a balance between agriculture and industry, with priority for improved agricultural productivity and comprehensive rural development. Industrialization was not abandoned, but special emphasis was given to its decentralization and its relationship to agriculture, including the production of farm tools and agricultural inputs and the processing of food and fiber outputs.

This type of development strategy seems likely to have major energy implications in the rural areas. The improvement of agricultural productivity through the introduction of high-yield varieties implies irrigation and drainage; application of chemical fertilizer; varying degrees of farm mechanization; and greatly increased transportation as subsistence farming is partly replaced by commercial farming (whether the crops are food or non-food). It also implies a pattern of physical infrastructure quite different from urban-centered industrialization, since agroindustries are promoted in widely spread towns. There are critical implications for rural electrification: Which fraction of the energy needs in farms and country towns should take the form of electricity? Should electricity be generated by decentralized or centralized technologies? Which fuel should be used? Which rate structures are appropriate?

The balanced strategy is intended not only to secure high overall rates of economic growth, but also to reduce income disparities between city and country, modern and traditional sectors, and social classes. This is especially the case when its agricultural component is focused on "integrated rural development"—a concept directed toward increased productivity and earning power for small farmers, sharecroppers, and landless rural laborers. The strategy is sometimes called "redistribution with growth."[10]

Basic Human Needs

Energy implications of the "basic needs" approach would depend on its more precise definition and the country environment in which it is carried out. In middle-income developing countries, the concept is obviously complementary to other development strategies. It implies a combination of structural measures to secure broader participation in a generally satisfactory growth performance (in oversimplified terms to make the patterns of Brazil and Mexico more similar to those of Korea and Taiwan), along with some reorientation of public services (education, health, housing sites, potable water, urban transportation) toward the rural and urban poor and a guaranteed floor of nutritional adequacy. There is a large overlap with the balanced growth strategy and also with regional policies, since extreme poverty and deprivation tend to be geographically concentrated (a classic example is the Brazilian Northeast). In these cases, additional energy implications of a basic needs strategy might be discovered (a) in direct household uses and (b) in requirements for altered patterns of public services and perhaps for the production of basic wage goods.

It is in the poorer countries, especially of South Asia and tropical Africa, that a more fundamental shift is proposed toward the meeting of basic needs as the first stage in an altered development strategy.[11] In those cases, energy implications would appear to flow from the emphasis on highly decentralized and relatively self-sufficient sub-economies; on labor intensity in both industry and agriculture; on the supply of low-cost wage goods; and on the general availability of basic education and health services—all within the constraints of low average income levels.

Despite these definitional uncertainties, rough estimates can be made of the energy requirements involved in some aspects of the basic

[10] Taken from the book of that title by Hollis Chenery and coauthors, *Redistribution with Growth* (London, Oxford University Press, 1974).

[11] See, for example, Paul Streeten and S. J. Burki, "Basic Needs: Some Issues," *World Development* vol. 6, no. 3 (1978), pp. 411–421; Paul Streeten, "From Growth to Basic Needs," *Finance and Development* (Sept. 1979) pp. 28–31; and Montek Ahluwalia, Nicholas Carter, and Hollis Chenery, *Growth and Poverty in Developing Countries,* Staff Working Paper, No. 309 (Washington, D.C., World Bank, 1978).

needs philosophy. For example, on the admittedly simplistic supposition that a certain minimum of household energy for lighting, cooking, and other direct consumption is needed to satisfy basic human needs, the calculations confirm that the total requirements would be substantial.[12]

The issue may be somewhat academic, since no developing country has been willing to give overriding priority to meeting basic human needs, and developing-country representatives in international discussions have often opposed the strategy on the ground that it is a prescription for permanent backwardness. A more realistic assumption would anticipate the absorption of the basic needs objectives into broader development strategies rather than an outright replacement of one strategy by another.

Radical Alternatives—"Another Development"

There has been some discussion in recent years of more radical approaches, typified by the publications of the Dag Hammarskjöld Foundation under the heading "Another Development" and by the writings of Ivan Illich.[13] These schools of thought oppose modern industrial society in both its open-market and its centrally planned versions, calling for its rejection in favor of life in rural areas and villages with simple, labor-intensive technologies and tightly knit communities. Their literature emphasizes such meta-economic goals as self-reliance, self-limitation of wants, environmental harmony, solidarity, participation, cultural integrity, and social justice. There is an intellectual kinship with the Utopian Socialists of the early nineteenth century and with much of the anarchist literature since Kropotkin.[14] The rejection of material consumerism also reflects a kinship with the moral teachings of many fundamentalist religious movements.

[12] John Jankowski, "Future Energy Requirements for Implementing Basic Human Needs and Other Development Programs," draft (Washington, D.C., Center for Energy Policy Research of Resources for the Future, 1980).

[13] Dag Hammarskjöld Foundation, *What Now? Another Development*, Dag Hammarskjöld Report on Development and International Cooperation (Uppsala, Sweden, 1975); *Development Dialogue*, A Journal of International Development Cooperation published by the Dag Hammarskjöld Foundation (Uppsala, Sweden, 1976ff); Ivan Illich, *Tools for Conviviality* (New York, Harper & Row, 1973); see also *IFDA Dossier*, an occasional publication of the International Foundation for Development Alternatives (Nyon, Switzerland, 1978ff).

[14] P. A. Kropotkin, *Selected Writings on Anarchism and Revolution* (Cambridge, Mass., M.I.T. Press, 1975).

The counterpart in energy strategy has been most eloquently presented by Amory Lovins.[15] Although his work relies on arguments that solar, biomass, and other soft technologies make economic sense even for the present structure of society, his analysis emphasizes the compatibility of such "soft" technologies with societies structured in a more decentralized way, avoiding "high" technology and giving lower priority to material consumption. If such a society were to come into being, it would almost certainly be less energy intensive than any of the other development strategies already noted. The energy demand mix would also involve a smaller proportion of electricity and of transportation, and hence a smaller call on oil. However, since these radical alternatives have thus far been practiced only in few experimental enclave communities, the majority located in industrial countries, there is no empirical evidence on their energy implications.

Development Tactics

Energy considerations may be of particular relevance in making certain types of choices among development tactics within an overall development strategy: industrial policy, concerning the mix of products to be produced; technology policy, concerning the way products are produced; and regional policy, concerning the geographic distribution of production. Energy supply schemes can also be directly involved with development tactics, as in the case of rural electrification. In most cases, the influence of energy on these choices will be smaller than that of factor endowments and social values, but energy costs and availabilities are becoming increasingly important constraints.

Industrial Policy

Before the era of cheap oil, locations with ample domestic supplies of coal, such as Britain, Germany, and the Pittsburgh region of the United States, had an advantage in pursuing industrialization—which was highly energy intensive, especially relative to the agricultural technologies of the day. When oil supplies became easily transportable, securely available, and cheap, it was possible for countries without important domestic energy reserves to industrialize on the basis of

[15] Amory Lovins, *Soft Energy Paths: Toward a Durable Peace* (Cambridge, Mass., Ballinger Publishing Company, 1977).

Table 4-3. Energy Coefficients for Selected U.S. Commodities, 1972

Commodities	Input of energy in dollars for each dollar of delivery to final demand
Energy commodities:	
Petroleum refining and related industries	1.657
Electric, gas, water, and sanitary services	1.322
Coal mining	1.213
Crude petroleum and natural gas	1.073
Other commodities:	
Chemicals and selected chemical products	0.263
Plastics and synthetic materials	0.159
Primary iron and steel manufacturing	0.133
Paper and allied products (except containers)	0.120
Stone and clay products	0.119
Primary nonferrous metals manufacturing	0.109
Broad and narrow fabrics, yarn and thread mills	0.098
Glass and glass products	0.096
Food and kindred products	0.066
General industrial machinery and equipment	0.065
Household furniture	0.060
Apparel	0.060
Electric industrial equipment and apparatus	0.060
Household appliances	0.060
Engines and turbines	0.059
Construction and mining machinery	0.058
Motor vehicles and equipment	0.058
Optical, ophthalmic, and photographic equipment	0.047
Radio, TV, and communication equipment	0.040

Note: This table gives total requirements, direct and indirect. Each entry represents the output required, directly or indirectly, from the four energy industries—coal mining, crude petroleum and natural gas, petroleum refining and related industries, and electric, gas, water, and sanitary services—for each dollar of delivery to final demand of the commodities listed.

Source: Philip M. Ritz, "The Input-Output Structure of the U.S. Economy, 1972," U.S. Department of Commerce, Survey of Current Business vol. 59, no. 2 (February 1979) table 5, pp. 67–71.

factor endowments other than energy. The most famous example is Japan, which capitalized on its technological knowledge and skilled labor to develop into a major industrialized power, and even an important exporter of energy-intensive products (such as steel and petrochemicals), on the basis of imported oil. Hydro power and natural gas, even cheaper than oil, gave some countries an advantage in especially energy-intensive industries, such as aluminum processing and ferroalloys, even prior to 1973.

As the share of energy in total costs of production rises, it becomes increasingly advantageous to locate energy-intensive industries near nontransportable, low-cost energy sources. As table 4-3 shows, the energy coefficient (the dollar amount of energy used to produce each dollar of output) is quite small for many industries, but some are conspicuously large energy users: ore refining and metals processing (especially aluminum refining, electrolytic conversion processes, and iron and steel making), cement, pulp and paper, petrochemicals, and the energy industries themselves. For reasons of both cost and reliability of supply, countries without major energy resources will tend to move away from these energy-intensive industries in their future development strategies; indeed, some are already doing so, including Japan. The logical candidates for relocation of this type of industry are countries with low-cost energy in forms involving high transportation costs—the clearest examples being water power, abundant natural gas, coal remote from the seaboard, and possibly some geothermal sites. But this opportunity will vary greatly by product: importing clay to make cement, for example, is unlikely to prove economical, but importing alumina to make aluminum using low-cost energy may have more promise.[16]

For other kinds of manufacturing industry, energy costs are unlikely to have a decisive effect on location, since they are outweighed by other factors. In any case energy costs tend to rise in a similar fashion in most countries. In essence, planning for industrial development, even in an energy-expensive age, must consider all factors of production. There is, however, a range of important industries for which reliable electrical supply, free of frequent outages or fluctuations in voltage and frequency, is a technical necessity, constituting another direct link between energy policy and industrial policy.

Technology Policy

The development goals of a society and the economic environment determine the appropriate technological choices in a country. In Japan, for example, the existence of a highly skilled and educated labor force and the lack of domestic resources have dictated an appropriate technology strategy of export promotion of high quality products incorporating much skilled labor, gradually augmented by increasing

[16] Michael Roemer, "Resource-Based Industrialization in the Developing Countries," *Journal of Development Economics* vol. 6 (1979) p. 176.

quantities of capital.[17] In China, the availability of cheap but unskilled labor and the priority accorded to food production led to a technological package which stressed agricultural output, first through labor-intensive infrastructure projects and later through manufacture in rural areas of productivity-raising agricultural inputs.

The critical element in the concept of appropriate technology in developing countries, an element that affects all combinations of productive factors, including energy use, is a higher ratio of labor to capital than would result from applying industrial-country technology. Do more labor-intensive activities use less energy than more capital-intensive ones? Intuitively, the answer would appear to be yes, since capital and energy seem to be complementary. But in industrial countries, a debate is currently raging over whether energy indeed complements or substitutes for capital. There is some indication that higher energy prices may lead to a replacement of both energy and capital by labor in the short run, but that capital replaces energy in the longer term, as more thermally efficient (and expensive) machines are adopted.[18] An important empirical question is whether, given different relative factor prices in developing countries, labor rather than capital might not be substituted for energy in the long term as well as the short. To be compatible with development, the labor substitution would have to avoid a loss in productivity; otherwise the shift would be a simple case of economic retrogression. In this field there is a clear need for technological research by developing countries, since labor-short industrial countries are unlikely to develop energy-conserving technologies of a labor-intensive type "appropriate" to developing countries.

What are the likely energy effects of increased labor intensity in production processes? The most labor-intensive processes—those that use no machines—use less commercial energy than the most capital-intensive. But it is less clear whether intermediate processes, which are probably most appropriate in terms both of using labor and increasing productivity, invariably use less energy per unit of output than the more capital-intensive ones. And even if efficient core processes are necessarily capital intensive, ancillary processes can be

[17] Hugh Patrick and Henry Rosovsky, eds., *Asia's New Giant: How the Japanese Economy Works* (Washington, D.C., The Brookings Institution, 1976) pp. 15–19, 105–125.
[18] Raymond J. Kopp and V. Kerry Smith, "Capital–Energy Complementarity: Further Evidence," RFF Discussion Paper D43 (Washington, D.C., Resources for the Future, October 1978).

made more labor intensive by "unpackaging" the components. These changes, such as transporting materials by hand or wheelbarrow instead of belts or forklifts, are likely to save energy as well.

Appropriate technology may be relevant to the supply of energy as well as the demand, by making a good fit to the factor proportions, labor skills, and material resources available in developing countries. The issue of labor intensity of alternative forms of energy supply is of special interest. For example, the effects of many renewable (or soft) energy sources on employment have been claimed to be positive. One study has suggested that the introduction of solar heating in the United States might cause a net increase in employment, primarily because of added small-scale construction work.[19] On the other hand, an increase in labor demands for solar energy may require a higher level of skills than is ordinarily believed. Moreover, if employment generation is to be considered a justification for choosing a relatively high-cost energy source, the possibility of providing jobs at lesser expense by other means should also be considered. One recent rough calculation shows an important employment-generating effect for fuel ethanol from sugar, as compared with alternatives such as hydrocarbon synfuels or methanol.[20] The higher costs of the ethanol option, however, would result in implied subsidies of about $2,000 a year to maintain each job created.

One major thrust of the search for lower cost alternatives to petroleum has been to increase the efficiency of use of traditional fuels in rural areas, through the addition of capital. The potential gains are so large that these kinds of improvement may lead to a net decrease in the overall capital–output ratio. Biogas plants, for example, produce more fuel and fertilizer than burning dung directly; solar cookers use direct solar energy; and improved wood stoves can increase the useful energy received from a quantity of wood by as much as tenfold. Because of their initial capital cost, these "appropriate" energy technologies—in the sense of using local and potentially renewable resources—may not be available to the poor, who cannot afford even the small monetary outlays involved. But in such cases, it might be

[19] Joseph C. Bruggink, "Macroeconomic Effects of Decentralized Energy Systems," in "Small-Scale Energy Technologies," draft (Washington, D.C., Center for Energy Policy Research of Resources for the Future, 1978) pp. 18-98–18-121.

[20] Alan Poole, "A Working Paper on Ethanol and Methanol as Alternatives for Petroleum Substitution in Brazil," draft (São Paulo, Instituto de Fisica da Universidade de São Paulo, August 1979) p. 161.

more accurate to describe as "inappropriate" the institutional arrangements that fail to encourage efficient investments, rather than the energy technologies.

Regional Policies

In many countries, regional disparities in incomes, employment opportunities, and social conditions are large and persistent and are not spontaneously overcome through internal migration. One aspect of regional policy is the promotion of greater equality among regions. Another is the decentralization of production and population in smaller centers in rural areas instead of the concentration of activities in one or a few large metropolises. The diseconomies of scale in terms of pollution, overcrowding, and the costs of urban services are evident in places such as Mexico City, where metropolitan area population has been estimated at 12 million already and still growing rapidly. The apparent remedies are discrimination in favor of the poorer regions and smaller cities in the provision of physical infrastructure and social services, and special incentives for industrial investment, tourism, or other generators of income and employment. Yet the industrial-country examples of the Italian South and Appalachia, as well as the unsuccessful attempts of Mexican, Brazilian, Nigerian, and other developing-country authorities to discourage or redirect rural–urban migration, show how difficult such policies are to implement.

Energy policy—especially but not limited to electrification—may play an important role in the success or failure of regional strategies, because of its impact on the location of productive activities. Evidence from the United States suggests that energy availabilities and costs have had a considerable influence on the location of economic activity.[21] Although there is as yet little empirical evidence to substantiate this finding on a global basis, similar effects are to be expected. Certain industries require particular forms of energy—a steel industry needs coal, for example—and virtually any kind of manufacturing will require electricity. Therefore, local availability of energy will be a critical "minimum" in a policy of industrial decentralization.

Costs of infrastructure, including energy, are considerably lower

[21] Irving Hoch, "The Role of Energy in the Regional Distribution of Economic Activity," paper presented at the Conference on Balanced National Growth and Regional Change, Lyndon B. Johnson School of Public Affairs, the University of Texas, Austin, Texas, September 1977, pp. 7–15.

per unit if demand is concentrated in towns, although not in cities beyond an optimum size. Thus, subsidies may be necessary to equalize production costs in rural and urban areas and to encourage decentralized development. Equalizing electricity costs and achieving early island-wide electrification appears to have been one factor in the successful decentralization of industry in Taiwan, for example; and particular attention has been paid by the Las Gaviotas Rural Development Center in Colombia to the development of low-cost energy and industrial production technologies to encourage settlement of the Orinoquia region.

Finally, transportation costs, in which energy figures prominently, will influence regional location of production in two ways: locating near energy supplies will reduce costs of transporting bulky fuels, such as coal; while locating near large markets for outputs (usually urban areas) will reduce costs of transporting final products for sale. Which kind of regional distribution of production will use more energy in transportation is uncertain. The modal mix between road and railroad or coastal shipping freight transport will of course be important: smaller activities are more likely to use truck transport, which is especially energy intensive, and in many developing countries a rail network has never been implanted. It has been speculated that one of the reasons why transport takes a larger share of developing- than of industrial-country energy use, is that enterprises are small in scale and scattered throughout urban areas.[22] On the other hand, a major purpose of decentralization of industrial activity in China has been to avoid excessive transportation, particularly important in a country where roads and other infrastructure are extremely poor.

Rural Electrification Policy

Rural electrification is an especially significant, and controversial, aspect of development-related energy supply policies. Substantial resources have been devoted by developing countries to rural electrification for both economic and social reasons—an estimated $10 billion up to 1971 in the non-Communist regions, with larger amounts expected in the following ten years.[23] The stated purposes are to stimulate

[22] Robert Nathans and Philip F. Palmedo, "Energy Planning and Management in Developing Countries: Thoughts Concerning a Conceptual Framework" (Upton, N.Y., Brookhaven National Laboratory, 1977) p. 22.

[23] World Bank, *Rural Electrification,* a World Bank Paper (Washington, D.C., October 1975) p. 17.

increased agricultural productivity and output through irrigation and mechanization, to foster the growth of rural industries, and to raise the living standards of rural people. Rural electrification is subsidized extensively.

Historically, the use of electricity, even more than commercial energy as a whole, has been closely associated with rising incomes and productivity.[24] Today, developing countries with higher per capita incomes typically consume more electricity per capita and also devote more investment resources to rural electrification than do poorer countries.[25] Nevertheless, the direction of causation in the relationship between electricity and rural economic development has not been well established.

Use of electricity in the rural areas of developing countries is very low compared to industrial countries or to their own urban areas. Geographic coverage is limited, and direct users within electrified areas are usually a small percentage of the population. Coverage and quantities used are higher in Latin America than in Asia, with Africa at the low end. Sectorally, residential use accounts for one-third to one-half of the total in most projects, and productive uses—industrial, commercial, and irrigation—make up most of the remainder.

In general, larger and more advanced localities are more electrified than smaller ones and tend to be better able to reap the benefits of electrification of households, agriculture, and industry. The household users in rural areas are largely the better off, although there is evidence that the poor also value electricity and in some cases are willing to allocate a high proportion of their income to its use. Amounts consumed are low in any event, with the predominant uses being lighting and ironing. Appliance ownership, which largely determines electricity consumption, correlates closely with income.

The most significant potential for economic development through rural electrification lies in agriculture and industry. Electrification can be important in raising agricultural output where irrigation requires pumping, although diesel-powered pumps can also provide this service, with cost-competitiveness being highly site specific. Increased agricultural output requires other factor inputs in conjunction with water,

[24] N. B. Guyol, *The World Electric Power Industry* (Berkeley, Calif., University of California Press, 1969) pp. 33–42.

[25] Alan M. Strout, "The Future of Nuclear Power in the Developing Countries," Report No. MIT-EL-77-006-WP (Cambridge, Mass., Massachusetts Institute of Technology, 1977), p. 14.

such as fertilizer and seeds, farming skills, and accessible capital and product markets. Other agricultural uses of electricity include manufacture of farm tools and other production inputs and processing of food and fiber products.

Rural electrification can also play an important role in stimulating rural industry where appropriate infrastructure and other inputs are available. Small motive units are most conveniently powered by electricity. Potential small-scale industries in particular are likely to find it preferable to draw their power from a public source, whether centralized or decentralized, rather than generating electricity for their own use. Slight differences in fuel costs are unlikely to outweigh other cost considerations, except in very closely balanced locational decisions by potential manufacturers, owing to the small part of energy in total costs. But the availability of electricity for industrial activities can be crucial in the establishment of rural industries.

Some of the reasons often given in favor of rural electrification are the indirect benefits expected to flow from the introduction of a major modernizing catalyst into an area. These are difficult to measure, and little effort has been made to measure them. It is in any case difficult to separate the effects of electrification from other aspects of economic development that often accompany it. In assessing these arguments, it should be borne in mind that some of the indirect benefits might be more effectively or economically achieved through other means.

The vital question of costs of electricity supply for rural consumption tends to be highly specific to each country and site. These costs are relevant both to choices between electrical and nonelectrical forms of energy supply and centralized or decentralized electrification. Distribution costs for rural service are generally high per kilowatt hour because of the low density of usage. But it makes a vast difference whether a high-tension grid already exists or is to be constructed mainly for urban and industrial use, with excess generating capacity, so that rural electrification needs only the addition of transformers and rural feeder lines. Generating costs, likewise, will be heavily affected by the presence or absence of low-cost primary energy sources, especially hydropower, surplus gas, or cheap waste biomass. For many of the productive services in small towns or on farms, the relevant competition is direct diesel power (for example, for irrigation pumping) or diesel-generated electricity. As oil prices increase, centrally generated electricity from non-oil sources will become relatively more attractive, and oil-fired central station generation will tend to lose its competitive advantage. The general question of fuel alternatives for generating electricity is discussed in chapter 8.

Energy Self-Sufficiency Policies

The shock of oil price increases and related uncertainties about assured continuity of supply have raised the question in all oil-importing countries of how energy imports might be minimized, if possible to the point of achieving complete energy self-sufficiency. All other things being equal, it is entirely understandable that every nation would like to obtain its energy supplies entirely from within its own borders. Energy is a vital component of all kinds of economic activity; security of supply is therefore enormously important; and no one likes to have a critical resource subject to controls by governments or organizations beyond his reach.

But all other things are never equal. Depending on the resource endowment of the country concerned, energy self-sufficiency may be so costly that its pursuit does more damage to economic development than a continuation of oil imports, perhaps on a reduced level, or the partial replacement of oil imports by coal or other forms of lower cost imports. Here again, as in other aspects of energy policy making, it is essential to deal with the problem in the context of overall development strategy, rather than as an isolated energy sector.

Because of the speed of recent oil price increases, there are many situations where substitute fuels or conservation measures are already fully competitive but are not yet adopted because of insufficient application of capital, institutional inertia, or the effects of energy pricing subsidies which insulate the consumer from the real cost of imported oil to the national economy. In such cases, the direction of policy remedies is fairly clear, although politically often difficult to implement; in particular, alternative means may be indicated to soften the impact of higher prices on low-income users. Beyond pricing oil at full cost, the practical issue for energy policy makers comes down to the extent and form of subsidies and regulations to encourage indigenous energy supplies or to other means of conserving imported oil. Such measures are usually advocated on grounds of energy security, reduced pressure on the balance of payments, or energy independence more generally.

It is useful to give separate consideration to security considerations in the narrow sense of protection against supply interruptions, as distinct from economic questions of relative cost. Diversification of forms of energy and sources of supply, together with emergency stockpiling, may be a more economical means of achieving an adequate degree of security than the complete replacement of imports. International institutional arrangements may also help, as illustrated by the emergency sharing agreement among industrial countries in the International Energy Agency. Some form of participation in such arrangements by oil-importing developing countries is an important issue for future international consideration.

On the side of economics, the basic question is whether the true national costs of increased self-sufficiency are less than the costs of available alternatives, such as expanding exports to pay for continuing oil imports at higher prices. In evaluating such costs, it is rational to recognize national objectives not internalized in the revenues of an enterprise responsible for indigenous energy, and to provide a subsidy accordingly. Such objectives include allowance for anticipated further increases in world oil prices, higher shadow prices for imports reflecting more realistic foreign exchange rates, lower shadow prices for labor when unemployment is high, carrying an "infant" energy industry through its early phases until the economies of scale and the learning curve make it fully competitive, or making use of idle land or agricultural wastes which are overpriced because of market imperfections.[26] But it is also important to give realistic weight to the opportunity costs of improving the balance of payments on the export side.

In the development of new forms of energy supply, moreover, international cooperation may often be more rewarding than the pursuit of energy autarky. In some cases, it will be advantageous to develop international river basins jointly for hydropower. If nuclear energy turns out to be the most desirable means of producing central station electricity, the economies of scale suggest that internationally shared reactors and spent fuel management might also be desirable. International technology interchange will in any case be indispensable to the efficient development of new energy sources. Alongside the effort to

[26] "Shadow prices" are used in project analyses to indicate economic costs to a society, reflecting real relative scarcities, rather than the financial costs actually paid by an enterprise, which may be affected by subsidies, unrealistic exchange rates, or other market distortions. For a more extended exposition of these matters, see Milton Russell, "Government Policies and Subsidies to Energy," paper presented at the UNITAR Conference on Long-Term Energy Resources, Montreal, November 26–December 7, 1979.

reduce dependence on increasingly scarce oil imports, therefore, there should be parallel efforts to identify other forms of international energy cooperation that may prove advantageous for the long term.

Conclusions

This review of the relationships between energy and development suggests that the linkage to economic growth is stronger in the developing than in the industrial countries and that there is no obvious alternative development strategy offering an easy escape from the constraints of higher energy costs. Some of the alternative strategies do have important implications for energy policy; that is especially true of the balanced strategy of agricultural and industrial growth, coupled with income redistribution and complementary regional and rural development policies.

Energy considerations are, however, of particular relevance in making certain types of choices within an overall development strategy—in defining industrial policy, technology policy, and regional policy. In most cases, the influence of energy will be smaller than the effects of factor endowments and social values, but energy costs and availabilities are becoming increasingly important as constraints on these choices.

APPENDIX 4-A
ENERGY DEMAND COMPARISONS BETWEEN IMPORT-SUBSTITUTING AND EXPORT-ORIENTED COUNTRIES

It is difficult to make quantitative comparisons between any of the development strategies detailed in the text; some qualitative points have already been sketched. However, it is possible to explore to some degree the difference between import-substituting and export-oriented industrial development. The classification of countries into these two categories is by no means clear. The situation is also clouded by the fact that explicit national development strategies are often not realized, so that an attempt at balanced growth may end up as import-substitution, or vice versa. However, table 4 A-1 attempts to trace the energy-intensity ratios for two groups of countries. From 1960 to 1970,

Table 4-A-1. Ratios of Industrial Production to GDP and Energy Consumption to GDP, for Selected Developing Countries
(absolute figures in metric tons coal equivalent)

Countries	1960 I/GDP[a] (%)	1960 E/GDP (tce)	1963 I/GDP (%)	1963 E[b]/GDP (tce)	1968 I/GDP (%)	1968 E/GDP (tce)	1970 I/GDP (%)	1970 E/GDP (tce)	1973 I/GDP (%)	1973 E/GDP (tce)	1976 I/GDP (%)	1976 E/GDP (tce)
Group A, import-substituting												
Brazil	20	580	20	580	21	630	21	590	18	580	21	590
Colombia	21	1,320	22	1,390	22	1,380	22	1,300	23	1,240	21	1,210
India	12	1,260	14	1,400	15	1,430	15	1,320	16	1,440	17	1,560
Mexico	27	910	27	810	28	810	29	880	30	910	31	950
Turkey	15	470	17	530	21	660	22	670	24	780	24	790
Group B, export-promoting												
Egypt	11	1,700	17	1,580	20	1,520	21	1,280	20	1,280	24	1,790
South Korea	13	1,250	16	1,760	23	1,900	25	2,170	32	1,930	48	1,670
Pakistan	12	1,340	16	1,440	19	1,550	17	1,380	19	1,200	16	1,150
Yugoslavia	31	1,280	34	1,290	37	1,310	38	1,320	42	1,420	41	1,350

Notes: I/GDP = share of industrial production in GDP. E/GDP = ratio of energy consumption to GDP.

Sources: World Bank Data Services, UN, *World Energy Supplies,* 1972–1976 and 1950–1974 (Washington, D.C., 1975 and 1977); United Nations, *Statistical Yearbook 1978* (New York, 1979).

[a] Industrial output ratios derived from the UN industrial production index and related to World Bank GDP data in 1975 dollars.

[b] Commercial energy consumption data from United Nations.

group A was characterized by some analysts as import-substituting, while group B was characterized as formerly import-substituting, but by the end of that period as more export oriented.[1] Although group B was in general more energy intensive, the data fluctuations obscure any systematic relationship that might exist between the development strategy and energy intensity. Therefore, the hypothesis that the energy intensity associated with import substitution is less than that for export-oriented industries remains only an interesting conjecture.

[1] See Hollis Chenery and Moises Syrquin, *Patterns of Development, 1950–1970* (Washington, D.C., Oxford University Press for the World Bank, 1975) p. 54; and J. B. Donges, "A Comparative Survey of Industrialization Policies in Fifteen Semi-Industrial Countries," *Weltwirtschaftliches Archiv*, vol. 112 (1976) pp. 626–659.

5

Energy Conservation

The two previous chapters suggested that if high rates of economic growth in developing countries are to be achieved, the demand for energy, particularly commercial energy, will continue strong. Differences in types of development strategy do not appear to alter this conclusion substantially. As energy prices are widely expected to continue rising, possibilities of increasing the efficiency with which energy is used become of major concern. Although the emphasis in this chapter is on saving energy, it should be remembered that inputs other than energy enter into decisions to produce and consume, and there will be many cases where maximizing energy efficiency in a technical sense may not be socially desirable. Nonetheless, it is important to identify the main energy-using, particularly oil-using, sectors and investigate in a preliminary way those possibilities of saving energy which can be achieved without major economic disruption.

Commercial Fuels

The main candidate for saving energy in the category of commercial fuels is oil—the fuel whose price has risen most, which is largely imported, and whose long-term future seems most in doubt. In those countries for which data are available,[1] transportation is the major market for oil, accounting typically for between one-third and one-half of total consumption. Industry and residential-commercial categories come next. "Nonenergy," that is, feedstocks for petrochemicals

[1] Brazil, India, Kenya, Mexico, Nigeria, Portugal, Thailand, and Turkey.

(although entirely composed of oil), does not often account for a large proportion of consumption. In general, there are few reliable data on energy consumption by the main end-use sectors in developing countries. Consequently, one of the first steps in assessing conservation potential in both the commercial and traditional sectors is data collection. This will involve drawing up energy balance sheets so that major end-use sectors and substitution possibilities can be identified. The traditional fuel sectors must not be forgotten. A good beginning for this exercise can be found in the energy balance sheets drawn up for some developing countries by the International Energy Agency working in cooperation with national agencies.

The Transportation Sector

The transportation sector is not only a major consumer, but is almost completely dependent on petroleum products, except in a few cases where railroads are still fueled with coal or are electrified on the basis of water power. There are only very limited possibilities in this sector for substituting other forms of fuel for petroleum products. Alcohol fuels such as ethanol and methanol produced from agricultural crops or wood are discussed in chapter 7. Conservation measures in this sector will involve either: (a) a less rapid rise in the size of the sector, (b) a change in the modal mix toward more energy-efficient forms of transportation, or (c) an improvement in the energy efficiency of the various transportation modes.

Differences and similarities among countries, including industrial as well as developing countries, can give some indication of the magnitude and feasibility of conservation options. With regard to the size of the transportation sector, national accounts data indicate that the value added by purchased transportation in the richer developing countries (6–8 percent of the gross domestic product [GDP]) is similar to the industrial countries. If the substantial passenger traffic in private cars were included, however, the industrial countries would almost certainly prove to have a larger modern transportation sector.

Given the strategic role of the passenger car in private transport, it is instructive to compare the number of passenger cars in relation to the level of income. Table 5-1 shows a wide range, from ten and fewer passenger cars per million dollars gross domestic product (GDP) in 1975 in Korea and India to forty in Brazil, Portugal, and Argentina. The case of Korea is especially pertinent. Here, despite a high degree

Table 5-1. Car Ownership Relative to Gross Domestic Product in Selected Countries, 1975
(cars per million 1975 $ GDP)

Countries	Car ownership relative to GDP
Industrial	
United States	69
France	45
West Germany	43
Japan	34
United Kingdom	61
Developing	
Argentina	41[a]
Brazil	40
Greece	21
India	9
Kenya	32
Korea	4
Mexico	30
Portugal	63
Thailand	18
Turkey	11

Sources: Car ownership from United Nations, *Statistical Yearbook 1977* (New York, 1978) p. 537; GDP from World Bank data services and from Organisation for European Co-operation and Development, *National Accounts of OECD Countries 1952–1977*, vol. 1 (Paris, 1979) p. 132.

[a] For 1974.

of industrialization and rapid economic growth leading to currently high living standards compared with other developing countries, car ownership relative to GDP is extremely low—one-half the level of the relatively poor countries such as India. At the other extreme, Brazil with an income per capita less than twice as high as that of Korea has ten times as many cars per million dollars of GDP—the same number as several European countries. The causes of such variations merit further investigation.

In freight transport, the amount of traffic will depend on the size of the country and on the amount and composition of industrial output. The largest countries tend to make the most intensive use of freight transport, and some industries such as iron and steel have a larger transport input than others. In addition, there is some evidence that energy intensity of freight varies with the stage of economic development. At times of rapid industrialization, the number of ton miles relative to GDP increases; indeed at some point this figure may become greater in developing than in industrial countries, whose typically more

Table 5-2. Energy Consumption by Road and Rail Transport Relative to Total Energy Consumed by the Transportation Sector, 1976

(percentage)

Country	Road	Rail
OECD (Europe)	80	5
United States	85	3
Japan	76	6
Argentina	84	11
Brazil	87	3
Colombia	87	—
Egypt	93	—
India	41	54
Indonesia	89	1
Kenya	47	32
Korea	53	3
Mexico	90	4
Nigeria	75	2
Thailand	40	22
Venezuela	91	—

Notes: Air and coastal transportation account for the remaining energy consumption by the transportation sector. Dashes = nil or negligible.

Source: International Energy Agency/Organisation for Economic Co-operation and Development, *Workshop on Energy Data of Developing Countries,* vol. II *Basic Energy Statistics and Energy Balances* (Paris, OECD, 1979); Organisation for Economic Co-operation and Development, *Energy Balances of OECD Countries, 1975–1977* (Paris, OECD, 1979).

regionally balanced economic structures lead to stable or even declining freight transport.

A major characteristic of any transportation system from the point of view of energy conservation potential is the modal split—that is, the shares accounted for by road, rail, water, or air. This is important because it takes as much as three or four times the energy to perform a passenger or ton mile in road transport as it does in rail, indicating an important source of oil savings if road transport, especially private passenger transport, can be curtailed.

The proportion of total transport energy consumed in road transport dominates in all countries. In several developing countries it is higher than in industrial countries (see table 5-2). This may be caused largely by historical reasons; many developing countries were laying down the elements of a modern transportation system in a postrailroad era of cheap oil. But it does mean that developing countries now have a transport modal split which is heavily energy intensive, even when compared with the industrial countries.

A final characteristic of the transport sector that can affect the amount of energy used is the energy efficiency of each mode. The mileage achieved by passenger cars varies widely from as low as 7 miles per gallon to as high as 40—all from standard production automobiles. Countries with similar numbers of passenger-miles, equally wedded to road rather than rail transport, could consume widely differing amounts of energy. It is not possible to generalize at this stage about the fuel efficiency of the car and bus fleet in different countries. Given the price of gasoline, typically higher in the developing countries than in the United States, although with some exceptions, a reasonably energy-efficient fleet must be assumed for the future. But there are factors which might offset the tendency. In developing countries, passenger car ownership is frequently concentrated among the rich, whose sensitivity to high gasoline prices may take second place to the desire for large automobiles. For freight transport, a large proportion of the truck fleet in some developing countries has gasoline engines rather than the more energy-efficient diesels.

With this background, what are the possibilities for future conservation possibilities? As incomes rise, the demand for private transport is likely to grow much more rapidly. A study of urban energy use in Mexico City indicates that of all categories, consumption in transportation rises the most rapidly as incomes rise.[2] Current use of gasoline for passenger cars is still quite small because of the narrow ownership of cars. But, as is indicated by experience in industrial countries, gasoline consumption for private cars could rise to very high levels indeed.[3]

Moderating the growth in use of passenger cars has two aspects. First, the number of passenger miles travelled can be constrained by a variety of means including the development of public transportation facilities and better design of urban growth. In developing areas with particularly high rates of urban growth, new urban areas are created in predictably short periods of time. This provides an opportunity for planning future city growth consistent with energy-efficient transport modes.

[2] Gordon McGranahan and Manuel Taylor, *Urban Energy Use Patterns in Developing Countries: A Preliminary Study of Mexico City* (Stony Brook, N.Y., The W. Averell Harriman College for Urban and Policy Sciences, State University of New York, 1977) p. 54.
[3] It is indicative that at present the United States consumes 187 tons of gasoline per million dollars GDP, while India consumes only 15 tons.

Even given existing urban patterns, which grew up under a cheap energy regime, much can be done to increase the energy efficiency of the passenger car fleet by taxation of large cars and gasoline or by measures regulating energy efficiency. As already indicated, these factors may not be of importance at present when passenger car ownership is very limited, but they will become of increasing importance as development proceeds.

Another possibility is to pay greater attention to the many forms of private and public transport other than the private automobile. These include jitneys, taxis, buses, and vans—all of which have higher load factors and are therefore more energy efficient. Despite this advantage, taxi transport in its various forms is frequently regulated in many municipalities in the direction of reducing rather than encouraging high load factors.

Other forms of transportation include the numerous variations of motorized bicycles and bicycles. If the demand for private transportation and the flexibility this brings with it can be met by modes other than the passenger car, energy problems will be easier to manage.

With regard to freight transportation, there seems some indication that freight intensity—after a point—may decline with economic growth. This point, which is associated with a stage of balanced regional growth, may be quite a long time in coming for many countries. For the poorer countries especially, the first stages of industrialization may lead to very rapid growth in freight transport, which seems inevitable if development is to proceed.

This means more attention to alternative freight transportation modes and their efficiencies if energy is to be economized. It is clear that railways are much more energy efficient than trucks but they also have major disadvantages—they are inflexible, they require ancillary trucking services (including more transshipment), and their construction is highly capital intensive. If a country, such as India, already possesses a substantial railroad system, upgrading and extension may attract some traffic from the roads. If countries have inadequate rail systems to begin with, it may not be realistic to consider the construction of new railways except in very special cases, such as deliveries of coal to electricity generating plants where a large predictable volume of freight can be counted upon. Transportation by ship could also be important for countries like Indonesia and the Philippines, island states like Fiji, countries with long coastlines like India and Brazil, and with inland waterways like Thailand, Bangladesh, and Egypt.

This leaves much of the burden of energy conservation in freight transportation on improving the efficiency of trucks. There may be considerable room for improvement here in converting to diesel engines. As in all poor countries, however, the initial cost of change may be prohibitive. Individual operators and companies may not feel it possible to trade in their existing stock of vehicles in order to achieve greater fuel efficiency. The other side of the coin, however, is the relatively fast rate of economic growth in developing countries leading to a more rapid dilution of older energy-inefficient equipment.

The Industrial Sector

The industrial sector is generally the major consumer of commercial fuels in developing countries and, next to transport, the largest consumer of liquid fuels. The share of industry in total energy consumption has been increasing in many countries over the past ten years, and the share of oil in the rising total has generally increased or remained stable. The general rule seems to be that for those countries which had no alternative domestic fossil fuel resources, the percentage share of oil increased as the industrial sector grew. In other countries, such as Nigeria and Thailand, the share of oil fell, usually because of increased use of domestic natural gas or coal.

The amount of energy consumed by industry depends on the size of the industrial sector, the concentration of energy-intensive industries within the sector, and the energy intensiveness of the different industrial processes. The size of the sector differs greatly from country to country (see table 5-3). Broadly speaking, its size, measured by the share of GDP represented by value added by industry, is low at low levels of income, rises as income rises, and then has a tendency to level off at the highest levels of income as service industries increase in size relative to manufacturing.[4] In the future as in the past, developing

[4] United Nations Industrial Development Organization, *World Industry Since 1960: Progress and Prospects,* Special Issue of the Industrial Development Survey for the Third General Conference of UNIDO (New York, United Nations, 1979). Although this pattern is corroborated reasonably well by statistical data, a note of caution is in order. It is extremely difficult to ensure consistency in industrial value-added data. The size of establishments covered varies from country to country and the line between the service and industrial sectors is frequently difficult to draw. For the countries that concern us here, however, these distinctions are probably of less importance and do not affect the conclusion that the rise in income and size of industrial sector are closely connected.

Table 5-3. Industrial Energy Consumption in Selected Countries, 1973

Countries	Energy consumption by industry[a] relative to GDP[b] (metric tons oil equiv. per million $ GDP)	Value added by manufacturing as share of GDP (percentage)	Industrial energy consumption per unit of value added in manufacturing (metric tons oil equivalent per million $ value added by manufacturing)
Industrialized			
United States	359	25	1,436
Japan	325	36	903
France	272	33	824
West Germany	330	41	805
Italy	306	32	956
United Kingdom	326	27	1,207
Developing			
Argentina	135	33	409
Brazil	143	19	753
Greece	212	18	1,178
India	192	13	1,477
Indonesia	24	9	267
Kenya	126	11	1,145
Mexico	288	23	1,252
Nigeria	28	7	400
Portugal	133	32	416
Thailand	119	17	700
Turkey	93	21	443
Venezuela	313	18	1,739

Sources: GDP derived from Irving B. Kravis, Alan B. Heston, and Robert Summers, "Real GDP per Capita from More Than One Hundred Countries," *Economic Journal* (London) vol. 88 (June 1978); value added from: United Nations, *Statistical Yearbooks 1975 and 1976* (New York, 1976 and 1977); Portugal, Turkey, Greece, and all industrialized countries' energy consumption from: Organisation for Economic Co-operation and Development, *Energy Balances 1973–1975* (Paris, 1977); other developing countries' energy consumption from: International Energy Agency/Organisation for Economic Co-operation and Development, Workshop on Energy Data of Developing Countries (Paris, OECD, 1979).

[a] Includes energy consumed by industry and in nonenergy uses.

[b] 1973 per capita GDP measured in international dollars are from Kravis and coauthors, "Real GDP." The dollars represent the extrapolated 1970 purchasing power parity of foreign currency to the U.S. dollar, so that one 1973 international dollar has the same *overall* purchasing power as one 1973 U.S. dollar. The per capita GDP of each country was then multiplied by its estimated 1973 population, which was obtained from World Bank, *Atlas* (Washington, D.C., 1974).

countries may be expected to plan on continued expansion of the industrial sector, so that major energy conservation potential is not likely to flow from a deliberate slowdown in overall industrial growth.

Even within the context of a larger industrial sector there remains the possibility of moderating energy use through a shift in the product mix, since different industries have varying energy intensities. As table 5-3 shows, there is considerable variation in the amount of energy consumed relative to industrial output, both between the industrial and developing countries and within the developing-country group. Past experience suggests a move from light to heavy, more energy-intensive industry in the early stages of development. But, given the large foreign exchange costs for oil, governments might be expected, through changes in planning targets and techniques, to take a more active interest in modifying the industrial mix.

Industrial energy conservation can also be achieved without changing the industrial product mix by using more energy-efficient processes. There are wide variations in the amounts of energy used to perform specific processes. The energy used in producing a ton of crude steel, for example, varies from 0.6 metric tons oil equivalent in the United Kingdom to 0.4 tons in Italy. India has one of the highest energy ratios, at about the UK level. In all cases, energy inputs per ton of steel have been declining for some time.[5]

These kinds of variations depend on the technology used. In many industries—steel and cement, for example—technology has become more energy efficient over time. This seems at first sight at odds with the fact of declining energy prices over the twenty years up to 1973—a situation which might have been expected to encourage increased energy intensity. In these industries, however, energy is a major element of total costs so that reduced energy consumption, even at a low price, would represent a significant decline in total costs.

As we have already seen in chapter 4, the developing countries with relatively small stocks of existing energy-using equipment are well situated to take advantage of the newest technology. The difficulty is, however, that this technology may be highly capital intensive. The ideal solution for many countries that have relatively abundant supplies of labor may lie in technologies which substitute labor for both capital and energy (see chapter 4). It is not clear as a practical matter whether such technological choices are available.

There are also possibilities that energy is being wasted in significant quantities owing to negligence born of long periods of cheap oil. If so,

[5] Joel Darmstadter, Joy Dunkerley, Jack Alterman, *How Industrial Societies Use Energy: A Comparative Analysis* (Baltimore, Johns Hopkins University Press for Resources for the Future, 1977).

then substantial savings—reportedly as much as 10 to 15 percent[6] over a relatively short time period—may be derived from an improvement in housekeeping procedures at little extra expenditure.

Finally, oil can be saved through the substitution of other fuels, depending on the availability of other domestic resources. Those countries with coal (India) or gas (Mexico and Nigeria) consume less oil in industry than the other countries. Given a pipeline system, there is little difficulty in substituting gas for oil; in many circumstances it is preferred as a cleaner fuel. About one-third of U.S. industrial energy consumption is provided by gas, for example.[7]

The substitution of coal for oil is another matter. The experience of the 1950s and 1960s indicated that industry in industrial countries moved away from coal as quickly as possible. Besides being increasingly expensive in these years, coal had other disadvantages. It was dirty, more inefficient in use, difficult to handle, and expensive to transport. Following the oil price rise, coal is in many countries much cheaper than oil on a straight heat content basis, but the other disadvantages remain. As things stand, without the introduction of improved coal-using technologies and quite forceful encouragement from the government, widespread conversion to coal is not likely. India and China, however, are examples of countries with large industrial sectors fueled almost entirely by coal. Another frequently mentioned possibility among domestic resources is biomass and hydroelectricity (see chapter 7). But many countries do not have domestic energy resources and must rely on imports of other fuels if they wish to substitute for oil. In these cases, much will depend on whether a major international trade in coal and gas develops.

All in all, if developing countries are to increase economic growth, the industrial sector will have to grow, and it is not clear how much of a choice will be available between pursuing greater or lesser energy-intensive industrial strategies. Furthermore, fuel substitution possibilities may be limited in many countries.

The Residential Sector

When the traditional fuels are included, the residential sector is frequently the largest in developing countries, accounting for a rela-

---

[6] P. R. Srinivasan of the National Productivity Council, India, to Joy Dunkerley, January 1979.

[7] International Energy Agency/Organisation for Economic Co-operation and Development, *Energy Balances of OECD Countries 1974–1978* (Paris, OECD, 1980).

tively small part of total oil consumption—perhaps 10 to 20 percent.[8] This is because much of the population lives in rural areas where traditional fuels (wood, charcoal, animal and crop wastes) are used for cooking and other domestic functions (see chapter 2). Small quantities of electricity and kerosine are used in rural areas but the bulk of petroleum products for household use is used by urban dwellers.

Residential energy consumption depends on the stock of appliances in the home and the efficiency of those appliances. Although furnaces are not necessary for most of the developing countries, which are situated in warmer areas, in Latin America much of the population is in mountain regions or the south temperate zone, and many cities are cold enough to require space heating for part of the year.

Ownership of other appliances exists even at very low levels of income, and rises rapidly as incomes increase. The typical sequence begins with a stove for cooking (using various fuels). Then, as income rises and electricity becomes available, electric light, a radio, an electric iron, a fan, and a refrigerator are acquired. Television ownership is still sparse but rapidly growing even among lower income groups—evidence of the high priority placed on television. So far, air conditioning units are rare in houses, although much used in office buildings. The spread of air conditioning could introduce a very dynamic element into the demand for energy, given the potential desire for cooling in tropical and semitropical areas. While appliance ownership in general is much lower than in industrial countries, the basic appliances are known and used in urban areas almost everywhere, and the stock can be expected to rise rapidly in the future.

The position of petroleum in this sector is of particular interest. Petroleum products are almost essential in transport and preferred in industry, but in the residential sector, kerosine is considered an "inferior" fuel once a certain level of income is achieved. At low levels of income, kerosine is usually preferred to wood and sometimes to charcoal, but with rising incomes, consumers shift to natural or bottled gas or electricity. Kerosine is therefore a transitional fuel, and from the point of view of oil conservation, there may be considerable potential for bypassing this stage and moving directly from charcoal to gas and electricity—assuming that there are non-oil resources for electrical generation. This may be easier said than done, since the use

[8] International Energy Agency/Organisation for Economic Co-operation and Development, *Workshop on Energy Data of Developing Countries*, Vol. II *Basic Energy Statistics and Energy Balances* (Paris, OECD, 1979).

of kerosine is deeply rooted in the energy consumption patterns of many countries.

Electricity Generation

One other major use of oil is in electricity generation. In the countries we consider here this sector is large, often ranking with industry in importance. The size of the generating sector varies greatly, with consumption of electricity in some developing countries five or six times the level in others, even taking different levels of income into account (see table 5-4). Given the heat losses involved in generation of electricity, one possible conservation approach would be to burn fuels directly rather than convert them first to electricity, in cases where the energy service required is heat rather than light or motor power. It is necessary, however, to consider the "systems" efficiency with which competing energy sources are used. To cite an example of special interest to developing countries: wood fuel for cooking in open fires has an efficiency of 5 to 10 percent, while the efficiency of cooking with electricity is almost 100 percent. Therefore, even taking into account the large heat losses sustained in generating that electricity, the overall system efficiency of electricity in cooking may be much higher than that of fuelwood.

The high capital intensity of electricity generation and the large economies of scale tend to invest electricity systems with an imperative of their own. Thus the desire to improve capacity factors leads to low prices for electricity at off-peak times, thus encouraging the expansion of demand into uses which might otherwise have been provided by direct burning of fuel. The large economies of scale in electricity generation have also contributed to its wide use by encouraging the construction of large distribution systems to match large power stations. In systems wholly dependent on oil as the primary fuel, however, direct fuel costs are now tending to dominate capital costs to the point where the traditional features of electric power system pricing and construction may be undergoing fundamental changes.

Traditional Fuels

The previous section considered conservation possibilities for commercial fuels. But as has frequently been emphasized in this study, especially in chapter 2, traditional fuels are a major part of the total

Table 5-4. Electricity Production Relative to Gross Domestic Product in Selected Countries, 1973

Countries	Electricity generation[a] (10^9 kWh)	Ratio of electricity production to GDP (1000 kWh per million $ GDP)[b]
Industrialized		
United States	1,965	1,508
Japan	470	1,110
France	174	738
West Germany	299	1,068
Italy	146	903
United Kingdom	282	1,216
Developing		
Algeria	3	231
Argentina	27	428
Bangladesh	1	90
Brazil	63	386
Chile	9	514
Colombia	11	459
Egypt	8	372
Greece	15	624
India	73	377
Indonesia	3	73
Iran	12	272
Jamaica	2	656
Kenya	1	145
Korea, Republic of	15	489
Mexico	37	428
Morocco	3	439
Nigeria	3	78
Pakistan	8	251
Philippines	13	537
Portugal	10	563
Saudi Arabia	1	90
Thailand	7	318
Turkey	12	278
Venezuela	16	588

Sources: Electricity generation from United Nations, *Statistical Yearbook 1975* (New York, 1976). GDP derived from Irving B. Kravis, Alan B. Heston, and Robert Summers, "Real GDP per Capita from More Than One Hundred Countries," *Economic Journal* (London) vol. 88 (June 1978) pp. 215–242.

[a] Gross generation of electricity.

[b] For methodology of GDP derivation, see footnote b of table 5-3.

energy supplies of many countries. The use of these fuels is concentrated in rural areas that are often entirely dependent on traditional fuels produced within the immediate locale.

The use of these fuels in rural areas is closely integrated with every aspect of economic life in such a way as to make the end-use distinctions applying to commercial fuels—such as industry and household uses—of limited analytical use. Energy in the form of fuelwood, charcoal, and crop and animal wastes is both a by-product of rural activities and an input into rural activities, either in the form of animal and crop wastes used as fertilizer in the growing of food, or in the form of fuelwood for food preparation.

The main characteristic of both traditional fuels and draft animals is their great inefficiency in use. Wood and crop wastes are typically used with efficiencies not exceeding 10 percent and draft animals with an efficiency of about 6 percent.[9] These facts appear to open up major possibilities for improving energy efficiency. These possibilities are especially important as most rural areas of the developing world will continue relying heavily on traditional fuels and draft animals for their energy supplies for some time to come. Rising prices of commercial fuels are likely to slow their adoption in rural areas. But there is also great pressure on supplies of traditional fuels in many parts of the world as population expands.

From the technical point of view, there is little doubt that major efficiencies in energy use of traditional fuels could be achieved. Take for example, cooking, which accounts for most of the energy consumption in rural areas. Many forms of improved cookers exist, including solar cookers and improved wood burning stoves. As noted in chapter 2, total fuelwood consumption in the developing world is about 262 million metric tons oil equivalent annually, of which about 80 percent is used for cooking. If efficiencies were doubled—which given their present low level is not an unreasonable target—about 100 million metric tons oil equivalent (or 2 million barrels per day) of this biomass could be saved annually.

How To Achieve Conservation

Even taking into account rising consumption associated with increasing economic development, there are important opportunities to conserve energy in both the traditional and commercial sectors. Whether it will be possible to take advantage of these opportunities depends upon a

[9] National Council for Applied Economic Research, *Energy Demand in Greater Bombay* (New Delhi, Asia Publishing House, 1975).

number of considerations. The first is the cost of conserving energy. While the calculation will vary according to the end-use and particular circumstances of each consumer, in general for wide ranges and uses, conservation is reckoned to be significantly cheaper than the provision of extra supplies at current prices, and to have a much shorter lead time than most supply options. But these substantial cost advantages do not in themselves guarantee that conservation will take place. Prices of many forms of energy—kerosine, diesel fuel, coal, electricity—are frequently heavily subsidized, diminishing the incentive to conserve. In addition, consumers, particularly small consumers, may not have access to the necessary capital or may not be aware of the available opportunities. These considerations apply to consumers of both commercial and traditional fuels. The existence of major cultural, social, and institutional obstacles (see chapter 9) may also impede the adoption of otherwise attractive conservation possibilities.

Given these obstacles, it is all the more important to establish clear incentives to conserve. The nature of these incentives will vary according to the degree of central planning as opposed to autonomous decision making. In some countries, allocation is achieved almost entirely by physical controls. But in many of the developing countries, although there may be important elements of physical controls, some reliance is placed on pricing as an allocating device outside the centrally planned areas. The most effective single incentive to use energy more efficiently is for it to become more expensive. Underpriced electricity and petroleum products encourage consumers to use more energy— not the signals most governments really want to give. A general rule is that energy will be used more efficiently if each category of user pays the real replacement cost of the energy used. Even the present high market value for petroleum products does not necessarily represent possible long-term price trends and oil security costs. An extra tax or tariff on imported petroleum products or similar energy sources may therefore be indicated.

In addition to reforms of energy pricing, other measures may be needed to secure conservation objectives. Such measures include loans and subsidies to consumers, regulation of the energy efficiency of major energy-using appliances, and rationing. Whatever measures are chosen to back up higher prices, care should be taken to minimize the additional burden on an already strained bureaucracy. Administrative simplicity should be emphasized, modifying existing procedures rather than instituting wholly new ones.

Serious equity problems are raised however by the prospect of higher energy prices, especially for the household and transportation needs of lower income groups. For most industrial countries, these problems are handled better through income maintenance and other welfare arrangements rather than special energy pricing for the poor. In many developing countries lacking social security systems, however, price manipulation and subsidies may be the only available way to provide relief even though they benefit others along with the poor. Subsidies will not vanish overnight, and special measures may be needed to minimize perverse effects on energy supply and use. For example, some countries discriminate between different petroleum products by subsidizing kerosine and diesel fuel at low levels and taxing gasoline heavily. This can lead to the diversion of kerosine to use in place of gasoline or diesel fuel.

There is clearly a conflict in pricing policy between economic efficiency and equity. This conflict will not be resolved easily, and compromises will have to be made. It seems essential, however, for budgetary as well as efficiency reasons, for greater emphasis to be laid in the future on prices which reflect replacement costs of energy.

6

Fossil and Uranium Energy Resources

The Changing Problem

The energy supply problem means finding an energy resource—fossil or renewable—and adapting it to some needed energy end-use by using a convenient technology. The problems encountered in using both existing and new technologies are manifold, and they have been explored in a vast amount of literature on the subject. Therefore, we emphasize here only technologies in a few key areas of special interest to planners in developing countries (see chapter 8). Applying these technologies to practical situations may not be easy, especially in conditions where there are shortages of credit, skilled labor, and other important factors of production. These problems are discussed in chapter 9.

We begin, however, with the resources themselves; nonrenewable resources are treated in this chapter and renewable in chapter 7. We emphasize the resource problem here, because it has had little comprehensive treatment from the point of view of the developing nations, while types of technology have been treated extensively elsewhere. Moreover, the continuing upward pressure on oil prices has made the question of availability of economically feasible energy resources a key piece of the energy policy puzzle. We therefore examine the general outlook for supplies of fuels and other energy inputs in developing areas in enough detail to elucidate the general possibilities for the developing world.

To analyze the resource question definitively, it would be very desirable to be able to characterize resources by costs of extraction and processing. Although some of these costs exist, the data for most resources—with the prime exception of uranium—are usually not categorized by levels of extraction costs. In spite of this, in the context of rising prices of oil, estimates of resources considered "economically recoverable" by analysts in government and private industry are very important for energy policy decisions, even if the cost characterizations are necessarily imperfect.

Increasing domestic energy supplies is one of the priority tasks for energy policy makers in developing countries, and fossil and uranium energy resources, together with water power, are now the backbone of commercial energy supply in the world. They will almost surely remain so over the next few decades. Although most countries must take into account the continued need for importing oil over that period, energy planning in developing areas must take advantage of any possibilities of finding new oil, gas, and oil shale, or of discovering domestic coal or importing it from abroad. Depending on nuclear policy, new uranium deposits may also be important for some countries.

A resource of any kind is valuable only if it can be put to practical use. This means that many financial and institutional problems must be solved to turn oil in the ground into motor fuel. While the emphasis here is on the basic problems of geology that determine whether a resource exists at all in commercially viable concentrations, the political and economic aspects of the necessary institutions are also noted.

Oil and Gas

Finding new national oil or gas reserves is one obvious cure to the energy crisis for many developing nations. While this cure may not be perfect—Nigeria and even Mexico and Venezuela being good examples of continuing problems in the midst of energy abundance—the need for liquid fuels of high energy density, like petroleum, seems to be of continuing importance in any kind of reasonable energy scenario for the developing world.

The role of new petroleum exploration and development naturally is very dependent on the current price of oil in the world market, and therefore on the level of extraction costs that can be tolerated. Because of the many uncertainties involved, we have not attempted to assess the role of higher cost oil in any detail. However, the recent history of rising oil prices must be taken as a factor tending toward general optimism in seeking new petroleum resources in developing areas.

There are two contradictory aspects in the outlook for finding new oil and gas reserves in many developing areas. On the positive side, it is evident that many countries have not been adequately investigated for oil and gas. The negative side is that most—but not all—experts seem to believe that the amount of oil ultimately recoverable in the world is only between 230 to 320 billion metric tons oil equivalent (btoe), about 1.6 to 2.3 trillion barrels.[1]

The sources of this pessimism will be examined below, but we can immediately see how it provides a simple basis in arithmetic logic for the oil crisis when we compare these figures with oil consumption of about 3 btoe per year at the present time—and increasing every year. And of the 230 to 320 btoe, not all is still available: of the total of ultimately recoverable reserves, about 50 btoe have already been consumed. Furthermore, out of the 180 to 270 btoe remaining, 90 are in the form of presently identifiable reserves in existing fields, and of the remaining 90 to 180 btoe, it is expected that 60 to 100 will be found in as yet undiscovered traps (formations) in known oil fields. Therefore, this all leaves only about 30 to 80 btoe of oil that one can expect to be produced from entirely new discoveries.[2]

The situation is somewhat the same for gas reserves. Gas may be associated with oil fields or found in independent, unassociated fields. One estimate[3] has put the amounts of future additions to gas reserves from unassociated fields at roughly the same quantity as those from oil fields. It has been estimated that in total some 60 btoe of proven world reserves exist, plus some 190 btoe of as yet undiscovered gas.[4] It is an unfortunate fact that, under present economic constraints, much of this valuable gas is being flared and thus wasted.

A ray of hope may lie in the unconventional petroleum reserves such as tar sands and heavy oils. Heavy oils are often obtainable by quasiconventional recovery techniques such as steam injection and are

[1] Richard Nehring, "Giant Oil Fields and World Oil Resources" (Santa Monica, Calif., Rand Corporation, June 1978) pp. X–XI; Exxon Corporation, "Exploration in Developing Countries" (New York, June 1978) chart 1.

[2] Nehring, "Giant Oil Fields," pp. x–xi.

[3] R. L. Whiting, "Gas From Conventional Gas Fields," in R. F. Meyer, ed., The Future Supply of Nature-Made Petroleum and Gas (New York, Pergamon Press, 1977) p. 292.

[4] World Energy Conference, World Energy Resources 1985–2020, Executive summaries of reports on resources, conservation and demand to the Conservation Commission of the World Energy Conference (New York, IPC Science and Technology Press, 1978) p. 54.

virtually as useful as lighter petroleum; tar sands often require a different, mining-type technology, but the product is compatible with much of the standard petroleum technology. Exploration efforts to find reserves of these fuels are still at a rather modest level. Estimates have been made of about 23 btoe from tar sands and heavy oil. However, perhaps only a tenth or so of those are believed to be extractable using present technologies, and recent production estimates for the year 2000 have been put at only 46 million metric tons oil equivalent (mtoe).[5]

The techniques for constructing such global oil estimates are of more than academic interest: examination reveals good reasons for further pessimism for developing countries' energy prospects. Oil tends not to be found in random locations; most oil in large quantities is usually found within "provinces" (sedimentary basins) that can be identified on the basis of general geological knowledge. Of the approximately 600 of these basins that are known to exist, there has been some drilling in about 400 of them, with commercial production resulting from about 160 out of the 400, and with perhaps 40 more that could possibly be commercially productive in the future. That leaves 200 basins essentially unexplored. From the results achieved so far, one would then expect on an *a priori* basis that some 100 new basins might be productive of oil. Unfortunately, a great deal of this hypothetical oil will inevitably lie in basins that are in circumpolar areas or that are found in the deep ocean—in areas that may be impractical environments for drilling and transportation.[6]

Extrapolating from past results—at least given the present understanding of the nature of oil discoveries—may lead to even more negative conclusions for new prospects in developing countries that are not now producing oil. The raw numbers for basin statistics necessarily fail to take into account the "lumpy" and generally distinctive character of the distribution of present reserves and consequently the dim outlook for future reserves of very large deposits, even within large sedimentary basins or prospective oil-bearing provinces. The existing oil provinces in the Middle East and in western Siberia appear to possess unusually large deposits; they contain well

    [5] Ibid., p. 15.
    [6] Michael T. Halbouty, "Acceleration in Global Exploration Requirement for Survival," *The American Association for Petroleum Geologists Bulletin* vol. 62, no. 5 (May 1978) p. 746; Exxon, "Exploration," chart 1.

over 50 percent of the world's present oil reserves.[7] Some experts believe it is unlikely that the special characteristics of these basins—sediments of Tertiary and Mesozoic age characterized by a certain type of geology—would be likely to occur again. Furthermore, most of the 43 percent of the remaining reserves in the earth's crust are restricted to so-called "giant" fields in only 64 other sedimentary basins. Indeed, judging by past experience, we would be unusually fortunate to find even half as much oil as now exists in those 64 basins in the 200 or so basins still unexplored—not a comfortably large amount at present levels of demand.[8]

This familiar if gloomy picture of the world's resources of fluid hydrocarbons is quite misleading, however, in one crucial aspect: it concentrates on large reservoirs. Even though most of the world's oil may be in such reservoirs, substantial reserves must remain to be found in small and medium-size fields. Therefore, the oil exhaustion scenario should not discourage developing-area energy planners, but should only serve as a caution against excessive optimism. Obviously, for a country such as Burundi, reportedly using about 0.016 million metric tons oil equivalent per year, oil or gas finds that would be modest on a global scale could be of immense importance in its national energy balance. Furthermore, even if the total amounts of oil found in small reservoirs turn out to be much less than amounts needed for long-term energy self-sufficiency, they could still go a long way toward easing the transition from fossil fuels to renewable resources for some developing countries.

A glance at one estimate (in table 6-1) of the ultimately recoverable resources in geographical sections of the world heavily dominated by developing economies shows some of the possibilities: as seen in the table, the amounts of energy at stake may be quite significant in size. Naturally, that table does not tell the whole story. Areas such as Latin

---

[7] H. D. Klemme, "World Oil and Gas Reserves from Analysis of Giant Fields and Basins (Provinces)," in R. F. Meyer, ed., *The Future Supply of Nature-Made Petroleum and Gas* (New York, Pergamon Press, 1977) p. 250.

[8] It is possible that there is a deeper geological reason for this apparent "bad luck" that most of the oil exploration in really promising areas has already taken place. One speculation involves the theory of continental drift and the supposition that all major oil finds were once a single geographical zone—the postulated position of the continental plates during the Triassic period before continents had drifted apart. This area, 800 to 1,000 miles wide, contains 85 percent of the known oil reserves, including the seven largest oil-producing provinces, and also most of the unconventional resources such as tar sands and oil shale (Nehring, "Giant Oil Fields," p. 39).

Table 6-1. Fluid Hydrocarbon Reserves and Resources for Developing Countries, by Region
(million metric tons oil equivalent)

Region	Oil		Gas		Total	
	Proved reserves	Ultimately recoverable resources	Proved reserves	Ultimately recoverable resources	Proved reserves	Ultimately recoverable resources
Latin America & the Caribbean	770	3,570	560	1,600	1,320	5,200
East Africa	70	460	20	70	90	530
West Africa	120	2,270	2	160	110	2,440
South Asia	300	1,100	670	1,460	970	2,560
East Asia & the Pacific	140	1,300	440	1,530	580	2,800
Total	1,400	8,700	1,700	4,850	3,100	13,530

Note: Regions include only developing nations and are defined as follows:

Latin America and the Caribbean is the area south of the Rio Grande.

East Africa and West Africa are divided from each other by the Congo River and from southern Africa by the Zambezi.

South Asia consists of the countries lying between the Khyber Pass and the Salween River.

East Asia and the Pacific ranges from Thailand to Korea to Western Samoa.

Source: R. Vedavalli, *Petroleum and Gas in Non-OPEC Developing Countries,* Staff Working Paper No. 289 (World Bank, Washington, D.C., 1978) tables 3 and 4.

America, which have fairly large amounts of both proven and ultimately recoverable reserves, will tend to have those reserves associated with already discovered producing fields. Nevertheless, exploration is still worthwhile for many countries that are not now important oil producers.

What are the prospects for carrying out efficient exploration programs in individual developing areas? The sedimentary basins where oil is found can be or have been identified with relative ease. Can exploratory oil drilling in these basins quickly determine whether or not oil exists? The evidence from past exploration is somewhat reassuring. "Super-giant" fields (of about 700 mtoe potential) have always been discovered relatively early in the exploration of a province.[9] Furthermore, all giant fields (over 70 mtoe potential) have been found in large geologic "traps" that are detectable by surface means, and technical knowledge has improved so much over the years that such traps are extremely likely to be discovered. Indeed, the good prospects for new discoveries in promising basins in developing countries may be somewhat masked by comparing raw statistics on success of exploration in the past, since older exploratory efforts, especially in the United States, were often not efficient compared to modern techniques.[10] Furthermore, within each basin it is usually necessary to drill only a few wildcat wells in order to establish the largest oil-bearing structures.[11]

Finding smaller fields may be more difficult. But a compensating advantage is that more of them may exist waiting to be discovered. Exploring for these smaller fields may well be the most promising approach for many developing areas. We can hope that the progress made in recent years in exploration technology will also facilitate the finding and exploitation of such fields.

New exploration should probably also benefit from a more sophisticated approach to geological structure and the detection of traps. Petroleum appears to originate in organic-rich shale or carbonate

[9] Nehring, "Giant Oil Fields," p. 63.
[10] Ibid., p. 56n.
[11] Exxon, "Exploration," p. 2. A minority view (B. F. Grossling, "The Petroleum Exploration Challenge with Respect to Developing Nations," in R. F. Meyer, ed., *The Future Supply of Nature-Made Petroleum and Gas* [New York, Pergamon Press, 1977]) holds that global oil prospects are much brighter. Others maintain that this optimistic outlook depends on assuming many more basins and very large outputs per square kilometer (L. F. Ivanhoe, "Petroleum Prospects of non-OPEC LDCs," *Oil and Gas Journal* [August 27, 1979]), e.g., a hypothesis characteristic of U.S. experience involving small basins but perhaps unlikely as a general exploration rule (Nehring, "Giant Oil Fields," p. 71).

Figure 6-1. Potential Oil-Bearing Areas in South America. Derived from H. D. Klemme, "World Oil and Gas Reserves from Analysis of Giant Fields and Basins (Provinces)," in R. F. Meyer, ed., *The Future Supply of Nature-Made Petroleum and Gas* (New York, Pergamon Press, 1977).

"source" rocks.[12] After it is formed it commonly flows into other porous "reservoir" rock. Sometimes, otherwise promising sedimentary basins are deficient in rocks of one or both sorts, as has apparently happened in certain basins in West Africa.[13] Most experts assume that oil comes from organic deposits, and therefore has to be protected from exposure to oxygen in order not to break down into other

[12] J. D. Haun, "Occurrence of Petroleum," in R. F. Meyer, ed., *The Future Supply of Nature-Made Petroleum and Gas* (New York, Pergamon Press, 1977) p. 46.

[13] Exxon, "Exploration," p. 2.

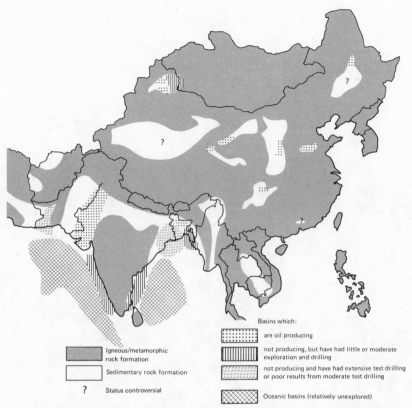

Basins which:

are oil producing

not producing, but have had little or moderate exploration and drilling

not producing and have had extensive test drilling or poor results from moderate test drilling

Oceanic basins (relatively unexplored)

Igneous/metamorphic rock formation

Sedimentary rock formation

? Status controversial

Figure 6-2. Potential Oil-Bearing Areas in Asia. Derived from H. D. Klemme, "World Oil and Gas Reserves from Analysis of Giant Fields and Basins (Provinces)," in R. F. Meyer, ed., *The Future Supply of Nature-Made Petroleum and Gas* (New York, Pergamon Press, 1977).

products.[14] In accordance with this, the petroleum must be trapped by harder rock strata in order to keep it from dissipating through the surface strata in the form of natural gas or other products.

As mentioned above, certain common types of traps often associated with large oil finds are readily detectable. However, there are other classes of traps that are not so easily detectable on the surface and have not been associated as often with oil finds in the past. These "stratigraphic" traps (and associated types) could be a potential source of oil finds in areas where the usual types of traps ("structural") have not previously been detected.

Detailed prospects for individual developing countries depend, of course, on country-by-country exploration. Maps in figures 6-1 through 6-4 show prospective oil basins in some regions and indicate the extent

[14] Haun, "Occurrence of Petroleum," pp. 45–54.

Figure 6-3. Potential Oil-Bearing Areas in Southeast Asia. Derived from H. D. Klemme, "World Oil and Gas Reserves from Analysis of Giant Fields and Basins (Provinces)," in R. F. Meyer, ed., *The Future Supply of Nature-Made Petroleum and Gas* (New York, Pergamon Press, 1977).

Legend:

- Igneous/metamorphic rock formation
- Sedimentary rock formation
- Basins which:
  - are oil producing
  - not producing, but have had little or moderate exploration and drilling
  - not producing and have had extensive test drilling or poor results from moderate test drilling
- Oceanic basins (relatively unexplored)

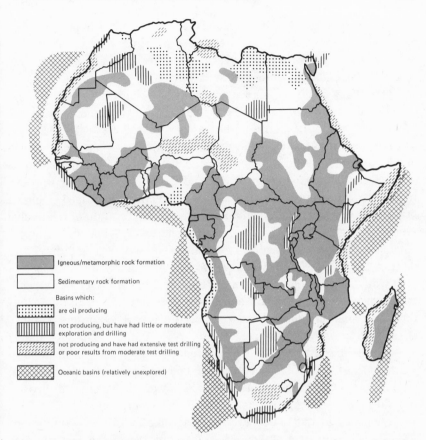

Figure 6-4. Sedimentary Basins and Potential Oil-Bearing Areas in Africa. Derived from H. D. Klemme, "World Oil and Gas Reserves from Analysis of Giant Fields and Basins (Provinces)," in R. F. Meyer, ed., *The Future Supply of Nature-Made Petroleum and Gas* (New York, Pergamon Press, 1977).

of success of the exploration. Recent exploration in Mexico and adjoining areas in Guatemala has been very successful. On the other hand, other efforts in the Central American and Caribbean region have proved disappointing, such as offshore drilling in Cuba. In South America, Ecuador and Peru have shown major promise in the Amazon area, while drilling in Tertiary rock in Colombia has not lived up to hopes.[15] In the meantime, many large surface traps remain undrilled in various sites in Argentina, and the Malvinas basin may have a major potential. In Asia, outside of the Middle East, the Bay of Bengal,

[15] A. A. Meyerhoff, "Where Major Onshore Reserves Will Be Found," *Oil and Gas Journal* vol. 75, no. 35 (August 1977) pp. 132–145.

onshore Burma, the Andaman Islands, and the South China Sea have been pointed out recently as areas of promise.[16] Several important areas in Afghanistan have also yet to be tested, while exploration in Pakistan has been characterized as neglected.[17]

For Africa, figure 6-4 shows some of the sedimentary basins that could hold oil. The areas already drilled—successfully and unsuccessfully—are shown. It should be noted that some believe that Tunisia has not been fully developed, and that several new giant fields could be discovered in North Africa, especially Algeria.[18] In sub-Saharan Africa, Chad shows some promise, and the Niger Delta could prove to be a "super province."[19] In Zaire, conditions are like some in China, in which difficult terrain could complicate the exploration of relatively young continental beds. The large Etosha basin lying in Namibia, Angola, and Botswana appears promising,[20] and parts of the East African coast, especially Somalia, are relatively untouched.[21]

It would appear that the estimates in table 6-1 give sufficient promise of significant reserves of oil and gas to repay future exploration by developing nations. Indeed, such exploration is proceeding and is being stimulated by recent initiatives of the World Bank.[22] While it cannot be claimed that all institutional problems are solved, the developing world has not been totally neglected even in the past: 50 percent of all exploratory wells in the decade 1967–1976 were reported as drilled in non-OPEC developing nations.[23] These wells were sunk in 71 out of 113 countries in that category, while seismic and other exploration was carried on in an additional 22. To be sure, the vast majority (5,416 out of 6,501) of these wells were sunk in 16 countries where significant discoveries had already been made before 1967. Of the other 55 countries, encouraging finds were made in only 25.[24] In fact, the absolute number of wells drilled in non-OPEC developing countries fell to 324 in 1977, compared with 609 in 1972. In addition, the 1976

[16] U.S. Central Intelligence Agency, National Foreign Assessment Center, *The World Oil Market in the Years Ahead,* a research paper (Washington, D.C., August 1979) p. 28 (available from National Technical Information Service, 79-103-70).

[17] Meyerhoff, "Major Onshore Reserves," p.137.

[18] Ibid., p. 135.

[19] Nehring, "Giant Oil Fields."

[20] Meyerhoff, "Major Onshore Reserves," p. 135.

[21] U.S. Central Intelligence Agency, *The World Oil Market,* p. 28.

[22] World Bank, *Energy in Developing Countries* (Washington, D.C., 1980).

[23] Exxon, "Exploration," chart 6.

[24] Ibid., charts 7 and 9.

"density" of wells drilled was only 7 per thousand square miles in non-OPEC countries, compared with a world average of 109.[25]

It is not easy to sort out the significance of this real or apparent "drilling gap." However, one can note that the geological prospects of a developing country are only one of several factors affecting exploration decisions by producing companies.[26] A second main factor is political stability, while a third concern is obtaining an adequate return on capital. These kinds of factors will not impede exploration in some countries, but in many areas fears of expropriation, lack of confidence in contracts, or uncertainties about local pricing and taxing policies have a deadly effect on oil prospects. Furthermore, there is often a fundamental economic conflict between the profit goals of producing companies and the desire of many countries to develop any national oil resources no matter how small.

Little can be done about the problem of political stability in a general sense. But the other problems, involving relatively assured rates of return and discrepancies between national goals and petroleum company interests, can be attacked through institutional means. Where rates of return are now uncertain because of undecided government policies and apprehension over control of resources by outsiders, international organizations can help develop new institutional capabilities through advice and support to individual governments. Some organizations are in fact now trying to act as intermediaries for some developing countries in negotiating with the major oil companies. Others are helping developing countries to deal more effectively with oil-producing companies through training, contract design assistance, and "unpackaging" various facets of exploration and development that have traditionally been performed by one large firm.[27] Furthermore, the World Bank has recently been active in helping with national energy planning in developing areas, and even more cogently, with assistance in petroleum survey work, exploratory drilling, and project preparation for oil resource use.[28] In some cases, the financial participation of the World Bank or a regional development bank may reassure both the host government and the private partners that the terms are

---

[25] R. Vedavalli, *Petroleum and Gas in Non-OPEC Developing Countries: 1976–1985*, Staff Working Paper No. 289 (Washington, D.C., World Bank, 1978) p. 9.

[26] Ibid.

[27] David N. Smith and Louis T. Wells, Jr., *Negotiating Third World Mineral Agreements: Promises as Prologue* (Cambridge, Mass., Ballinger, 1975).

[28] Vedavalli, *Petroleum and Gas*, p. 28.

fair and thus reduce the result of arbitrary renegotiation or expropriation. All these programs cannot solve completely the institutional
problems. But they can help alleviate any efficiency or equity problems
generated by petroleum "drilling gaps" and so both help individual
developing countries and help to increase petroleum production in the
world as a whole.

Coal

One of the most tantalizing possibilities for responding to the energy
crisis—which has been principally an oil crisis—is to search out and
use increased quantities of coal and other solid hydrocarbons. To be
sure, coal has its problems as an energy source. It contains mineral
impurities or ash that cause difficulties both as a solid waste disposal
problem and in the form of microscopic particulates polluting the
atmosphere—to say nothing of the infamous atmospheric carbon
dioxide effect that it shares with other fossil fuels. Sulfur impurities in
coal also cause environmental problems, and the occupational health
problems, such as mining accidents, that are associated with its
production are of great concern in most coal-producing countries.
Since coal in its natural state cannot be used in many energy applications, notably in internal combustion engines, it is far from being a
perfect substitute for oil. Nevertheless, coal has three particular
advantages: (1) it is a familiar fuel that is widely used and was formerly
an even more important segment of the energy picture than it is today;
(2) it is usually cheaper than oil on an energy unit basis; and (3) it is
abundant, at least on a global scale.

These advantages have complexities attached. Even though coal is
familiar as a boiler fuel in the United States and Europe, recent severe
difficulties in trying to reconvert electric utilities from oil to coal in the
United States cast some doubt on the importance of familiarity versus
other problems, such as need for new environmental controls. On the
other hand, the technology does work and engineering expertise does
exist, both in industrial countries and in such developing countries as
India and China.

The apparently favorable price structure also requires comment. If
the laws of the marketplace work, coal prices might well be depressed
(in raw energy unit terms relative to oil) because coal is not a fully
adequate substitute for oil. Only to the extent that some uses of oil
remain unique would one expect that the price of coal would continue

to lag behind, as oil prices rise in what one fears will be an upward spiral. To be sure, some analysts believe that despite pressure from oil prices, competition within the coal industries should also serve to keep prices down over the foreseeable future.[29] If so, the usefulness of the coal resource in the energy picture will be greatly heightened.

The abundance of coal as a resource is also not a simple, straightforward concept. Table 6-2 and figure 6-5 indicate the extent of world coal reserves. The messages of table 6-2 are varied. In the first place, the amount of coal resources, about 7,000 btoe, is some thirty times the amount of oil available in the world. Unfortunately, many of these resources would not be economical to use at the present time, and the amount of currently exploitable "reserves" is only about 430 btoe; but this is still a very large quantity. A thought-provoking aspect of the table is that the non-centrally planned developing countries, although having 10 percent of the world's economically recoverable reserves and 6 percent of current production, have only 2 percent of the vaguer, more speculative category of "geological resources." In contrast to earlier efforts, the projections shown, which were derived from the World Energy Conference of 1977, include consistent criteria for geological and economic evaluation.[30] Nevertheless, some experts believe that the figures given may be unduly conservative.[31] Given the contrasting ratios of economic reserves to geologic reserves for the industrial and developing countries, the degree of conservatism could well be even more for the developing world. This conclusion is reinforced by the observation that exploration for coal, as opposed to oil, has not been carried out to any extent in many areas: the map in figure 6-5 shows many countries where no solid fuel production is reported, and where no estimates of reserves exist.

Several examples of the probable conservatism in making coal resource estimates can be mentioned, based on informal accounts reported since the last available official statistics were gathered. Significant deposits may exist in many African countries; for instance, reserves of soft coal in Nigeria and other deposits in Zimbabwe, Swaziland, and Botswana. The last country may have reserves of 700 btoe, ten times higher than the 70 btoe listed in the official statistics.

[29] World Bank, *Coal Development Potential and Prospects in the Developing Countries* (Washington, D.C., October 1979) pp. 9–10.

[30] World Energy Conference, *World Energy Resources, 1985–2020—Coal Resources* (New York, IPC Science and Technology Press, 1978) pp. 20–22.

[31] Ibid., p. 40.

Table 6-2 World Coal Resources

Coal resources	Geological resources		Technically & economically recoverable reserves		1977 Production	
	World	Developing countries	World	Developing countries	World	Developing countries
In billion metric tons coal equivalent	10,125	230.0	636.0	65.0	2.77	0.18
In billion metric tons oil equivalent	6,890	155.0	433.0	44.0	1.90	0.12
Percentage of world total	100	2.3	100	10.2	100	6.30
Reserves as percentage of resources	—	—	6.3	28.3	—	—

Note: See Annex 1 of source for precise definitions of "geological resources" and "technically & economically recoverable reserves." Dashes = not applicable.
Source: Adapted from World Bank, *Coal Development Potential and Prospects in the Developing Countries* (Washington, D.C., 1979) pp. 4-5.

Figure 6-5. World Coal Deposits. Adapted from World Energy Conference, *Survey of Energy Resources* (New York, The United States National Committee of the World Energy Conference, 1974) endpaper. Reprinted by permission of The United States National Committee of the World Energy Conference.

LEGEND

Coal deposits

Non-reporting countries
(Generally, non-reporting countries have little coal production, reserves, or resources)

Similarly, in Latin America, it is quite possible Colombia may have reserves of 38 billion metric tons (26 btoe), instead of 8 billion metric tons (5.4 btoe), and that Brazil may have as much as 35 billion metric tons (24 btoe), plus 60 billion metric tons of lignite (20 btoe), not 10 billion (3 btoe) as reflected in the totals in table 6-2. New sources of coal in Asia may include possibilities in Bangladesh, Taiwan, North and South Korea, the Philippines, and Indonesia.[32]

What are the possibilities for finding more coal on general geological grounds? While the biological character of the oil formation process is somewhat controversial, the wide association of coal deposits with fossil plants makes it clear that coal seams of significant size are formed from the accumulation of large amounts of dead vegetation. For the vegetation to avoid being oxidized and lost through natural decay, it is presumed that coal must be formed in conjunction with bodies of standing water, such as swamps. Therefore, while intervening geologic upheavals must be taken into account, coal is usually found associated with areas of low relief that could reasonably be related to such water bodies. Climate conditions at the time of formation would probably be temperate or subtropical; tropical forests would produce the required vegetation, but would also tend to decay quickly through bacterial action before the coal could be formed.

While the temperature and relief constraints may rule out many areas in developing countries, many other areas will likely turn out to be potential coal-bearing regions. Some clues could come from the presence of oil deposits, since often the sedimentary and climatic requirements for oil formation are also conducive to coal formation. However, most coal was formed in earlier geologic eras (obviously including the Carboniferous Era), while the largest oil deposits come from Mesozoic and Cenozoic strata.

Against that background, it would be surprising if much more exploration for coal did not occur in the near future. Coal has, of course, suffered from competition with cheap oil. But even so, coal exploration and development have also been held back in the past by other factors: lack of geological data and other preinvestment activities, lack of capital for coal mining facilities and transportation infrastructure, and lack of engineering and managerial expertise. But perhaps most important is the low priority many national planning organizations give to coal exploration. These institutional problems, however, are

[32] Ibid., pp. 24, 27.

not usually of an insuperable nature. Projected near-term capital requirements for mining itself are often not very high, ranging from $20 to $150 per metric ton of annual output, or from one to seven times typical values of the annual coal output.[33] To stimulate the collection of geological data and other exploration activities, the World Bank's current program plans to start two to four coal or lignite projects per year.[34] If coal development proceeds successfully, the pace of developing countries' efforts toward coal utilization will probably increase at even faster rates.

Some developing countries that could potentially produce coal might also be able to produce more coal than needed for their own purposes. Such nations could then be added to the list of possible suppliers of coal to hydrocarbon-deficient countries. For the latter, imported coal, while not a domestic resource, could be an option of great importance for many developing nations as long as the international coal market develops in a relatively competitive fashion (see appendix 6-A).

Shale Oil

In a chemical sense shale oil deposits fall someplace between oil and coal. The organic residues contained in the shale are primarily carbon and hydrogen, in ratios taken to average about 10 to 1.[35] Compared to petroleum, this carbon-to-hydrogen ratio is high (for gasoline, the ratio is about 5.3 to 1), and therefore the average heating value per metric ton of shale organics is lower than that of petroleum. Nevertheless, when the shale is crushed and heated in a retort to relatively high temperatures (500°C), useful liquids and gasses are produced, and the liquids can then be further refined into substitutes for petroleum products.

A disadvantage of shale oil extraction is that, although an old technology, it has not yet been tested out on a large scale under modern conditions. Furthermore, the average energy content of shale-containing deposits tends to be low relative to coal, so that conventional surface mining for shale can disturb five times as much land as coal for

[33] World Bank, *Coal Development*, pp. 18–21.

[34] Ibid. p. 24.

[35] Donald C. Duncan and Vernon C. Swanson, *Organic-Rich Shale of the United States and World Land Areas*, U.S. Department of the Interior, Geological Survey Circular 523 (1965) pp. 4ff.

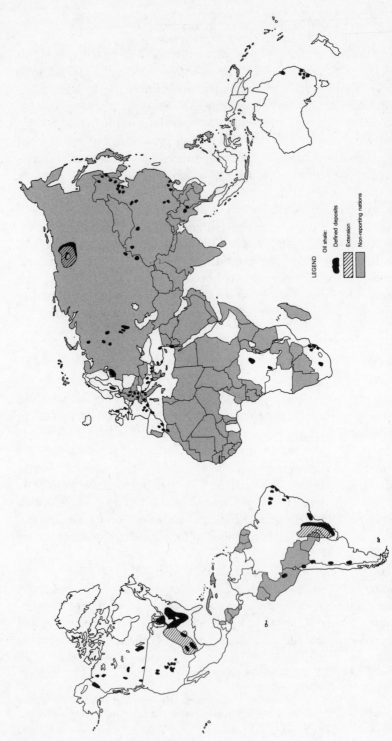

Figure 6-6. Oil Shale Resources. From World Energy Conference, *Survey of Energy Resources* (New York, The United States National Committee of the World Energy Conference, 1974) endpaper. Reprinted by permission of The United States National Committee of the World Energy Conference.

146

the same energy yield.[36] Water requirements can also pose a problem in arid or semiarid areas. Such environmental costs may perhaps be mitigated by newer and less tested *in situ* techniques. This more costly indirect underground mining would be somewhat analogous to oil drilling, but at a higher cost for shale oil.

Costs of shale in a U.S. context have been very roughly estimated at $170 to $300 per metric ton oil equivalent in 1975 dollars.[37] The lower end of this cost scale would be competitive at oil prices of $30 per barrel (in 1980 dollars).

The question of how much oil shale is available, and where, is of great interest. Earlier estimates show large identified global resources (at a certain minimal oil content) of 460,000 mtoe.[38] Estimates of those recoverable under present conditions indicate 27,000 mtoe.[39] Even so, resources in some parts of the developing world may be understated. Asia, for example, is shown with only 13 percent of the total, and this proportion may not be consistent with its large reserves of conventional hydrocarbons. Figure 6-6 reveals many intriguing large blank spaces in addition to evidence of shale deposits in countries such as Brazil, where there is serious interest in developing this potentially valuable resource.

Uranium

On the basis of past exploration and production, it is believed that there is little uranium in the category of "reasonably assured resources" located in the developing areas, at least as a share of the world total. Table 6-3 shows the amounts and proportions of uranium to be found in various areas that include many developing countries, versus

[36] Sam H. Schurr, Joel Darmstadter, Harry Perry, William Ramsay, and Milton Russell, *Energy in America's Future: The Choices Before Us* (Baltimore, Johns Hopkins University Press for Resources for the Future, 1979) p. 371.

[37] Conversion factors to change from dollars per metric ton of oil equivalent are $1/metric ton of oil equivalent (toe) is equal to $1.47/metric ton of coal equivalent (tce), or $0.0232/gigajoule (GJ), or $0.0245/million Btu. For example, fuel priced at $150/toe is equal to $150 × 1.47 = $220/tce, or $150 × 0.0232 = $3.48/GJ, or $150 × 0.0245 = $3.68/million Btu. (Schurr and coauthors, *Energy in America's Future*, p. 263).

[38] World Energy Conference, *Survey of Energy Resources, 1974* (New York, United States National Committee of the World Energy Conference 1974) p. 145.

[39] V. A. Ovcharenko, "Reassessment of Oil Shale Prospects." Paper presented at the UNITAR Conference on Long-Term Energy Resources, Montreal, Canada, December 5, 1979.

Table 6-3. Estimated World Resources of Uranium Recoverable at Costs up to $130 per kilogram, as of January 1977
(Excluding China, Eastern Europe, and the USSR)

Area	Reasonably assured resources		Estimated additional resources	
	Metric tons	Percentage of total world	Metric tons	Percentage of total world
World region (north)				
North America	825,000	38	1,709,000	79
Western Europe	389,300	18	95,400	4
Australia, New Zealand, & Japan	303,700	14	49,000	2
Total (north)	1,518,000	70	1,853,400	85
World region (south)				
Latin America	64,800	3	66,200	3
Middle East & North Africa	32,100	1	69,600	3
Africa, south of Sahara	544,000	25	162,900	7
East Asia	3,000	0	400	0
South Asia	29,800	1	23,700	1
Total (south) (incl. southern Africa)	673,700	30	322,800	15
Total world	2,191,700	100	2,176,200	100

Source: World Energy Conference, *World Energy Resources 1985–2020* (New York: IPC Science and Technology Press, 1978) p. 116. Reprinted by permission of the United States National Committee of the World Energy Conference.

amounts present in other parts of the world. In addition to these estimated reserves, a category of "estimated additional resources" that are somewhat less certain to be present is also shown in the table.

It can be seen that estimates of both types of resources, for example, for Latin America, are small as a share of world resources. Even in absolute terms, since a standard one gigawatt (GW) light water reactor will require some 200 tons of uranium to manufacture its yearly fuel load, the amount given for Latin American reserves is only equivalent to about 300 reactor years, or say ten reactors for a 30-year lifetime. This is not a large figure for an entire continent, even one featuring many countries in early stages of development. The same holds for Asia and, since most of the resources shown in Africa are located in South Africa and Namibia, also for most developing areas in Africa.

There are, however, several qualifications that should be taken into account in this pessimistic view of developing countries' uranium reserves. The cost of yellowcake (a mixture of uranium oxides) at present prices is a relatively small portion of the cost of nuclear fuel, and the fuel cost in turn a small portion of the cost of nuclear energy.[40] Furthermore, nuclear capital costs, although rising, are still low enough to make nuclear power competitive in many situations, especially where large generating stations are appropriate. Therefore, uranium costs as high as $260 per kilogram ($100 per pound of yellowcake)[41] could still be competitive. At such higher costs, one would expect much greater amounts of economically recoverable reserves. Based on the calculations made for a recent U.S. study, but probably with more general application,[42] one can estimate that costs of nuclear production of electricity under U.S. conditions would be as shown in table 6-4 for higher uranium prices. While fuel costs become dominant over nonfuel costs at between $130 and $260 per kilogram, costs of 6¢ per kilowatt hour (kWh) are possible even for extremely expensive uranium. Such electricity costs are not necessarily prohibitive for some electrical end-uses and in some fuel markets, especially in many developing countries that are poor in hydroelectric resources.

The other important factor to be remembered in qualifying the conventional gloomy outlook is that past mining of uranium has been

[40] Schurr and coauthors, *Energy in America's Future*, p. 283.
[41] World Energy Conference, *World Energy Resources*, p. 125.
[42] Center for Energy Studies, "Future Central Station Electric Power Generating Alternatives" (Austin, Tex., University of Texas for Resources for the Future, 1979) table 3.1-13.

Table 6-4. Nuclear Electricity Cost as a Function of Uranium Price
(1975 U.S. cents per kilowatt hour)

With uranium costs at: (dollars per kilogram)	Nuclear fuel costs	Electricity costs at a capital charge of:	
		10%	15%
46	0.55	1.7	2.1
130	0.9	2.0	2.5
260	1.4	2.5	3.0
520	2.5	3.6	4.0
1,040	4.6	5.7	6.1

Notes: The nonfuel costs included (but not shown) are: (1) Capital charges of 0.96 cents at 10 percent (1.44 cents at 15 percent) and (2) operating and maintenance costs of 0.15 cents. Fuel processing costs are 0.37 cents. Fuel costs include the category of "inventory costs" given in the source publication. To obtain uranium costs in terms of dollars per pound of U_3O_8 (yellowcake), multiply by 0.385; that is, \$130/Kg U = \$50/lb U_3O_8.

Source: Center for Energy Studies, "Future Central Station Electric Power Generating Alternatives" (Austin, Tex., University of Texas for Resources for the Future, 1979) table 3.1-13.

very restricted in geographic scope. Indeed, various estimates using crude mathematical or statistical models have reported that potential, that is, undiscovered, resources in the world may amount to between 80 to 280 million tons of uranium.[43] Such findings may or may not be correct. However, it is reasonable to compare known uranium resources to the known patterns of geological provinces in the world. On the basis of such a gross geological comparison alone, it would be surprising if North America, with only 17 percent of the non-centrally planned world land mass, really had over 58 percent of the total estimated known resources.[44]

This optimism can be backed up by considering the geological aspects of uranium exploration. Historically, uranium has been found in a wide variety of rock formations. Sandstone-type deposits and quartz-pebble conglomerates have been important types of deposits yielding uranium, as have certain types of veins.[45] However, other deposits, such as alum shales in Sweden, have been estimated as potentially large contributors. In addition, there are so-called unconventional sources of uranium, especially from phosphate deposits, certain copper ores, monazite produced in the heavy-mineral beach-sand industry, marine black (for example, "Chattanooga") shales,

[43] World Energy Conference, *World Energy Resources,* p. 122.
[44] Ibid.
[45] Ibid., p. 115.

above-average grade granites, coals and lignites.[46] Some recent speculations envisage that such uranium supplies could be large, by analogy with general patterns of trace element distribution in granite rocks.[47] Current conventional wisdom expects that costs for such sources will be higher than the $130 per kilogram range considered in table 6-3, and perhaps even higher than $260 per kilogram. However, the full potential of such resources has not been analyzed as a function of costs, and the cost barrier may not be fatal, as mentioned above. Supplies of uranium would be essentially limitless if uranium were to be extracted from sea water, but costs would be very high. The use of breeder reactors, it should be noted, would make the price of nuclear electricity quite insensitive to uranium costs.

In contrast to fossil fuels, then, uranium appears to occur in many different types of geographical locations. Indeed, since it is not connected with the existence of particular biological systems at particular geological epochs, it would be quite reasonable to speculate that its deposits should be more varied than those for fossil fuels. Of even more interest is that the number of types of geological situations in which uranium has been found has increased greatly within the past few years. Indeed, it has been noted that a significant amount of the radioactive gas, radon, comes from burning some types of coal because of the common occurrence of uranium in coal. (It has even been proposed that some coal ashheaps could be mined for their uranium content!)[48]

The search for uranium in new places and in new types of deposits can be even more optimistically regarded—at least in the long run—if local alternative costs of electricity someday rise to a high enough level to justify the use of possibly vast amounts of high-cost uranium. However, short-range planning must take into account the current softness of the uranium market and the prospect that higher prices for uranium may be some years in the offing; exploration should concentrate on high-grade formations for the foreseeable future.

While institutional problems in uranium utilization have yet to be subjected to as close an examination as they have been for hydrocarbons, it can be expected that the same type of conflicts between

[46] Ibid., p. 121.
[47] Kenneth S. Deffeyes and Ian S. MacGregor, "World Uranium Resources," *Scientific American* vol. 242, no. 1 (January 1980) pp. 66–87.
[48] William Ramsay, *Unpaid Costs of Electrical Energy* (Baltimore, Johns Hopkins University Press for Resources for the Future, 1979) p. 40.

commercial and national interests could arise. These problems should be muted, however, by the more indirect and less familiar role of uranium in energy usage and by the possibility that large high-grade deposits may be waiting to be discovered.

In view, therefore, of the very restricted amount of past exploration and production of uranium, and the fairly wide a priori possibilities for finding suitable types of geological deposits, it would seem that uranium exploration is a prime candidate for the attention of developing-area energy planners. However, this recommendation is in the domestic context only relevant if nuclear power is both cost effective in the usual financial sense and is acceptable from the point of view of safety, proliferation dangers, and other environmental problems. This is also the case if uranium is to be exported. These questions will be discussed in a later chapter.

Conclusions

Understanding the role of domestic supplies of mineral fuels—oil, gas, coal, uranium, and related substances—is crucial to the successful solution of energy problems in developing areas. Therefore, the geological prospects for these fuels are of great importance. All three of the mineral fuel types considered, fluid hydrocarbons, solid hydrocarbons, and uranium, promise good if often modest prospects for further finds in developing areas. While oil prospects may be the most restricted, in the sense that a great deal of oil exploration has already taken place—at least relative to other energy minerals—it is still the fuel of most immediate usefulness and flexibility. Institutional problems of finance and management can be key roadblocks, and various efforts are now being made to solve them. Heavier oils, tar sands, and shale oil may be found and utilized in significant quantities. Natural gas, which poses some use problems in most areas, will often be a by-product of oil exploration.

It seems quite likely that exploration for both coal and uranium has been inadequate in the past. Both these fuels have problems, some of which will be discussed below in chapter 8. But increased exploration efforts may pay handsome dividends in supplying energy needs for developing areas by earning valuable foreign exchange as export minerals. In addition importing coal could be a viable substitute for domestic supplies if a world coal market develops successfully from its present modest level.

APPENDIX 6-A
THE OPTION OF IMPORTING COAL

Even a country without indigenous coal may consider importing coal to replace higher priced oil imports. Although much attention is given to self-sufficiency in energy, most countries are in fact physically unable to provide for all of their energy needs from domestic sources at tolerable cost levels. As noted in chapter 4, partial reliance on imports can be far more economical and need not involve insecure supplies if the sources of supply are reasonably diversified and operating in competitive markets. For countries with ample fossil fuels, moreover, what they burn internally represents—whatever the pricing policy for the local market—an opportunity cost equal to world market prices.

The amount of coal available for export will depend on current prices and on government actions in coal-exporting countries such as the United States, South Africa, and Australia. However, one may expect the supply to be very responsive to demand factors in the present rather open state of the market, and additional countries such as Colombia and Botswana may enter the coal export field.

While prices of imported coal in developing countries will naturally reflect transportation costs, it should be noted that these are not necessarily forbidding if the consumption is near seaboard. Transport costs should be judged in relation to mine and production costs, which have been running $10 to $15 per metric ton for U.S. surface mines and $20 to $30 for underground mines; prices tend to be higher in Europe where mining is often subsidized.[1] While shipping costs vary, one model study showed costs of about $0.0008 per metric ton-kilometer.[2] For a 10,000 kilometer ocean voyage, such as to Japan from some U.S. ports, this would amount to $8 per metric ton transportation costs, a significant but not astronomical addition to the basic mining costs.[3]

[1] World Bank, *Coal Development*, p. 6.
[2] National Academy of Sciences, *Critical Issues in Coal Transportation, Proceedings of a Symposium* (Washington, D.C., 1979) pp. 60–61.
[3] These costs are for ocean transport by a 60,000 ton ship, and would be significantly smaller for a 120,000 ton vessel.

Of course, there must be added to these costs the higher (per. kilometer) costs of transport within the supplier country, other preparation costs, and costs in bringing the coal from the port to end-uses in the importing country. Costs for coal transport by railroad vary greatly, but one set of figures comes to 1.48 cents per metric ton-kilometer, or $7.40 per ton for a 500 kilometer train trip.[4] Much would depend on the availability of a heavy-duty railroad infrastructure. The total cost for an ocean shipment plus a long rail trip could then well mount up.

Importing coal could require establishing long-term contracts with supplier nations plus solving the same problems that expanded domestic use of coal would bring forth. That is, an infrastructure for coal-burning industries, especially utilities, must be established. Providing such facilities can be costly in both physical capital and scarce human skills, and only detailed study will show its desirability for use in any given developing area.

[4] Richard L. Gordon, *U.S. Coal and the Electric Power Industry* (Baltimore, Johns Hopkins University Press for Resources for the Future, 1975) p. 74.

7

Renewable Energy Resources

Renewable energy resources include direct solar energy and other newly prominent resources such as geothermal energy. They also include water for hydroelectric power, which is one of the most important sources of electrical energy in the world, and biomass—the wood, crops, and wastes that form the basis for the traditional energy sector in developing countries. For the distant future, as fossil and mineral fuels approach exhaustion, these renewable energy resources are the only hope—except possibly for nuclear breeds or fusion. But the relevant policy question is: how available are these resources for cost-effective use within the next few decades?

The number of renewable *possibilities* could constitute—and often has constituted—a formidable catalog of energy policy options. We are much more selective here in treating the resources and in chapter 8 in examining technologies. Wind resources, for example, are omitted here, and wind generators are considered only briefly below. We also do not treat other indirect solar sources such as wave power and ocean thermal energy conversion. Almost nothing is said about such exotic biomass alternatives as water hyacinths; readers may find brief descriptions of these in other sources.[1] In no way should such omissions be taken to indicate an assessment that such technologies will not work or will not be useful in the future; in fact, we believe that some of the more exotic options may be promising as future energy sources. But here our rationale has been to emphasize a selected group of technologies that are (1) sufficiently developed technically that they can be readily adopted by developing nations now or in the very near future

[1] See, for example, Sam Schurr, Joel Darmstadter, Harry Perry, William Ramsay, and Milton Russell, *Energy in America's Future: The Choices Before Us* (Baltimore, Johns Hopkins University Press for Resources for the Future, 1979) chapters 10 and 11.

and (2) common enough and sufficiently economical to use that they could contribute significant amounts of energy to national economies.

The preceding chapter showed that data are often woefully incomplete and that a great deal remains to be known about supplies of conventional energy resources. The situation is even more parlous with renewable resources, as indicated in chapter 2 on the traditional sector.

In fact, in many developing areas, the biggest thrust in present-day energy planning is often to gain an idea of the potential renewable resources available, whether biomass, solar, or geothermal. The need for such data should be evident from the brief review we give below of this problem. Inevitably, information gathering is costly, and some judgment should be exercised about the amount of labor and other resources to be expended on inventories of renewable energy sources. It should be evident that a careful study by energy analysts, ranking the most promising possibilities, should be used to set some priorities for data gathering in this area.

Hydroelectric Power

According to estimates made in the early 1970s by the World Energy Conference, the total amount of energy available per year from installed and "potentially installable" hydroelectric sources is about 820 million metric tons oil equivalent (mtoe) in electricity output—the equivalent of about 2,500 mtoe in fossil fuels used for generating electricity.[2] But present installed capacity produces an energy output of only 110 mtoe, or about 13 percent of the estimated potential.[3]

Although the estimated total is not large compared with future world energy needs, it does represent relatively large amounts of electricity —a particularly useful and flexible energy form. Presently installed hydroelectric capacity of 400,000 megawatts (Mw) accounts for about one-quarter of world electrical capacity.[4] Furthermore, the amounts of

---

[2] This is the energy corresponding to capturing the average flow of the watercourses included in the assessment for about half the hours in the year. The average flow measure will be used here, rather than measures involving more extreme flow requirements, such as a flow which will occur for 95 percent or more of the operating hours.

[3] World Energy Conference, *Survey of Energy Resources* (New York, The United States National Committee of the World Energy Conference 1974) p. 189. Original units in Terajoules (1 TJ = 10^{15} joules).

[4] Toby Gilsig, "The Development of Hydro-Electric Resources." Paper presented at the UNITAR Conference on Long-Term Energy Resources, Montreal, Canada, December 5, 1979, p. 2.

Table 7-1. Regional Summary of Potential Hydroelectric Resources and Recent Annual Production

(absolute figures in million metric tons oil equivalent)

Region	Potential annual energy[a]		Recent energy produced	Approximate percentage developed
	Fossil fuel input equiv.	Electrical output		
Africa	520	170	2.5	1.5
Asia (excl. Oceania)	680	220	16.5	7.5
Latin America and Caribbean	470	155	9.3	6.0

Source: Calculated from World Energy Conference, *Survey of Energy Resources* (New York, The United States National Committee of the World Energy Conference, 1974) pp. 187–189. Reprinted by permission of the United States National Committee of the World Energy Conference.

[a] Figures given are *net* electrical energy in units of million tons oil equivalent. For equivalents in kWh, multiply tons oil equivalent by 1.2×10^4 kWh; so Africa, for example, has a potential annual energy of 2.04×10^{12} kWh.

potential hydroelectricity still available in at least some developing nations are very large compared to any foreseeable projections of local needs. Table 7-1, for example, shows that for Africa only 1.5 percent of the hydro potential is now developed, while for Asia the total is only 7.5 percent and for Latin America 6 percent.

A possible hitch in this promising outlook for extending hydro power may be in the unknown—and in some cases undoubtedly forbidding— economics of existing or potential projects included in the totals listed as "installable" in table 7-2. Some of these economic problems will be discussed in chapter 8. In particular, transmission distances can be very large, with correspondingly high transmission costs and energy losses. On the other hand, older estimates of which projects are "installable" may not represent fully the economic outlook for hydro-electricity in the context of increasingly expensive thermal fuel sources. Moreover, recent interest in small-scale hydroelectric projects, including run-of-the-river projects, is not generally reflected in these figures.[5]

The calculations of installable capacity in the table, therefore, may have to be supplemented by considering how much more of the entire

---

[5] For example, it has been estimated that rehabilitating and expanding generators at larger existing hydroelectric facilities in the United States could increase total hydro-electric energy produced by 19 percent and that installing electrical generators of fewer than 5 megawatts at existing non-hydro dams could raise that to 48 percent (Institute for Water Resources, *Estimate of National Hydroelectric Power Potential at Existing Dams* [Alexandria, Va., U.S. Army Corps of Engineers, July 1977] p. 1).

Table 7-2. Potential Hydroelectric Resources and Recent Annual Production

Continent, region, or country	Installed and installable capacity[a]		Recent annual production	
	Power (megawatts)	Annual energy (terajoules)	Power (megawatts)	Annual energy (terajoules)
Africa				
Western Africa				
Nigeria	1,515	29,808	345	5,666
Ghana	1,615	55,984	948	9,515
Upper Volta	12,000	172,800	—	—
Mali	3,520	38,016	1	4
Ivory Coast	780	39,160	225	2,808
Sénégal	4,400	63,360	—	—
Guinea	6,400	92,160	20	90
Niger	9,600	103,680	—	—
Sierra Leone	3,000	43,200	—	—
Benin	1,792	25,805	—	—
Togo	480	6,912	2	18
Liberia	6,000	108,000	38	792
Mauritania	2,000	21,600	—	—
Portugese Guinea	120	1,728	—	—
Total, Western Africa	53,222	802,213	1,579	18,893
Eastern Africa				
Ethiopia	9,214	201,672	221	3,906
Tanzania	20,800	299,520	60	1,170
Kenya	13,440	193,536	71	1,148
Uganda	12,000	259,200	150	2,927
Mozambique	11,290	162,576	90	1,134
Malagasy	64,000	1,152,000	35	464
Zimbabwe	5,000	72,000	705	20,240
Malawi	100	1,440	40	511
Zambia	3,834	55,210	756	10,800

Rwanda	—	—	22	306
Somalia	240	2,592	—	—
Mauritius	80	1,152	26	184
Réunion	82	1,264	25	263
Total, Eastern Africa	140,080	2,402,162	2,201	43,053
Middle Africa				
Zaire	132,000	2,376,000	597	10,292
Cameroon	22,960	413,280	193	4,100
Angola	9,664	173,950	234	2,880
Chad	3,440	37,152		
Central African Republic	11,040	158,976	12	158
Congo, P. R. of	9,040	162,720	15	198
Gabon	17,520	315,360	—	—
Equatorial Guinea	2,400	43,200	1	7
Sao Tome and Principe	—		2	7
Total, Middle Africa	208,064	3,680,638	1,104	17,642
Northern Africa				
Egypt	3,800	54,000	2,448	19,260
Sudan	16,000	172,800	27	324
Morocco	975	10,800	300	5,393
Algeria	4,800	51,840	286	1,159
Tunisia	29	180	29	180
Libya	160	1,728	—	—
Total, Northern Africa	25,764	291,348	3,090	26,316
Southern Africa				
South Africa	4,600	30,766	180	2,700
Lesotho	490	9,360	—	—
Southwest Africa	1,200	12,960	—	—
Botswana	2,984	32,227	—	—
Swaziland	700	10,080	—	—
Total, Southern Africa	9,974	95,393	180	2,700
Total Africa	437,104	7,271,754	8,154	108,594

(continued)

159

Table 7-2 (continued)

Continent, region, or country	Installed and installable capacity[a]		Recent annual production	
	Power (megawatts)	Annual energy (terajoules)	Power (megawatts)	Annual energy (terajoules)
Asia				
East Asia				
China, P.R. of	330,000	4,752,000	10,000	137,050
Korea, Rep. of	5,514	35,730	621	6,498
Korea, D. P. R. of	2,000	28,800	3,600	41,760
Taiwan	1,632	19,019	1,131	13,630
Total, East Asia	339,146	4,835,549	14,752	198,938
Japan	49,592	468,014	19,897	296,172
Middle South Asia				
India	70,000	1,008,000	6,750	137,160
Bangladesh	1,307	23,526	120	—
Pakistan	20,000	378,000	593	10,800
Iran	10,196	134,795	798	12,953
Afghanistan	6,000	64,800	185	1,152
Sri Lanka	1,180	16,992	195	3,002
Nepal	—	—	23	173
Total, Middle South Asia	108,683	1,626,113	8,664	165,240
Southeast Asia				
Indonesia	30,000	540,000	315	4,799
Philippines	7,504	70,542	641	9,288
Thailand	6,242	81,302	451	5,400
Burma	75,000	810,000	103	1,440
Vietnam, D.P.R. of	48,000	691,200	150	2,160
Vietnam, Rep. of	5,598	101,314	709	13,000
Malaysia	1,319	16,081	296	3,812
Total, Southeast Asia	173,663	2,310,440	2,665	39,899

Southwest Asia				
Turkey	15,200	235,084	876	9,004
Iraq	1,900	27,360	96	1,868
Saudi Arabia	900	9,720	—	—
Syria	—	—	16	216
Lebanon	—	—	246	3,020
Israel	120	1,296	—	—
Total, Southwest Asia	18,120	273,460	1,234	14,108
Total Asia	689,204	9,513,575	47,118	714,357
North America				
Total, Canada & U.S.	290,209	4,581,753	85,905	1,565,820
Middle America				
Mexico	20,344	357,696	3,559	53,935
Guatemala	1,176	21,168	96	1,152
El Salvador	900	16,200	108	1,782
Honduras	4,800	86,400	40	828
Nicaragua	3,600	64,800	57	1,152
Costa Rica	4,326	136,433	182	3,726
Panama	2,400	43,200	20	324
British Honduras	300	5,400	—	—
Canal Zone	—	—	47	1,166
Total, Middle America	37,846	731,297	4,109	64,065
Total, Caribbean	2,400	43,200	196	2,117
Total, North America	330,455	5,356,250	90,210	1,632,002
South America				
Tropical South America				
Brazil	90,240	1,869,397	10,484	219,043
Colombia	50,000	1,080,000	1,900	25,596
Peru	12,500	392,954	935	12,420
Venezuela	11,644	352,811	895	26,338
Ecuador	21,000	453,600	105	1,584

(continued)

Table 7.2 (continued)

Continent, region, or country	Installed and installable capacity[a]		Recent annual production	
	Power (megawatts)	Annual energy (terajoules)	Power (megawatts)	Annual energy (terajoules)
Bolivia	18,000	324,000	171	2,178
Guyana	12,000	259,000	—	—
Surinam	260	5,854	180	3,600
French Guiana	233	6,750	—	—
Total, Tropical South America	215,877	4,744,566	14,670	290,759
Temperate South America				
Argentina	48,120	687,600	2,374	19,761
Chile	15,780	31,896	1,454	20,376
Uruguay	2,512	34,186	230	3,780
Paraguay	6,000	108,000	45	504
Total, Temperate South America	72,412	1,148,746	4,103	44,421
Total South America	288,289	5,893,312	18,773	335,180
Oceania				
Australia	8,605	88,139	4,310	44,471
New Zealand	9,994	198,720	3,200	57,600
Papua-New Guinea	17,762	438,012	30	648
New Caledonia	139	2,333	68	1,289
Western Samoa	15	252	1	22
Total, Oceania	36,515	727,456	7,609	104,030
World Total	2,265,974	35,304,871	307,362	4,713,302

Note: The energy potentials are given as evaluated for yearly average streamflow. To convert to metric tons oil equivalent, multiply the terajoule quantity by 23.2. To convert to kilowatt hours, use $2.78 \times 10^5 \times$ the terajoule quantity. Dashes = not applicable or not available.
Source: World Energy Conference, Survey of Energy Resources (New York, The United States National Committee of the World Energy Conference, 1974) pp. 187–189. Reprinted by permission of the United States National Committee of the World Energy Conference.
[a] At average streamflow conditions.

water runoff—the "gross theoretical potential"—can be used.[6] According to recent calculations for one country, only about one-third of the gross hydroelectric potential was assessed as technically exploitable in a practical sense.[7] Nevertheless, for some countries listed in the table, for which gross theoretical capacities have been calculated, the difference between "gross theoretical" and "installable capacity" is very large; in the case of the Ivory Coast the gross theoretical value is 5.3 mtoe compared with 0.9 mtoe (or 39,000 terajoules) shown in table 7-2 for installed and installable.[8]

The oil crisis has proved a spur to reassessing prospects for new hydroelectric development in many countries with low utilization of capacity. Africa is an obvious low-usage region, but many countries in Asia, for example, New Guinea, Burma, and Thailand, have been the scene of recent plans for more hydroelectric power. The push toward development of hydroelectric power must of course face its own peculiar problems, even from a purely technical point of view. In the case of countries such as India, Chile, and Thailand, for example, the variability of river flows often makes it difficult to utilize highly capital-intensive hydroelectric facilities economically. Furthermore, uneven flows tend to make the usual environmental and land use problems associated with hydroelectric power even greater—especially direct flooding and destruction of habitats for local species of flora and fauna. Finally, sedimentation problems can often be serious.

On the other hand, the other usual benefits of hydroelectric facilities, such as streamflow control and provision of water for irrigation, continue to be very important in many areas. Indeed, most hydroelectric planning decisions will inevitably—and properly—be made in the context of water resource development for the totality of local needs.

The key point to be emphasized is that hydroelectric resources need to be reinvestigated thoroughly in the light of existing fossil alternative costs and of prospective higher costs in the future. Old project feasibility studies should be reexamined, and planners should be given help in implementing new surveys of the resource base.

[6] Gross theoretical potential can be calculated either by so-called hypsographic profiles—that is, by constructing general relations between rainfall and altitude—or by measuring flows versus altitudes directly along all major watercourses. In one study for Turkey, these methods yielded fairly similar results (U. Özis, "Maximal Development of Water Power Resources." Article 3.1-7 of the Tenth World Energy Conference Transactions, Istanbul, Turkey, September 19–23, 1977, p. 11).

[7] Ibid.

[8] World Energy Conference, Survey of Energy Resources, p. 171.

Direct Solar Energy

Contrary to the common impression, the so-called sunny climes are not always more favored for solar power than higher latitudes. Indeed, if the attenuating effects of the atmosphere did not have to be taken into account, the longer summer days nearer the poles would somewhat more than compensate for the increased slant of the sun's rays in calculating annual totals of solar energy delivered to the surface of the earth. Of course, long winter nights make for an awkward seasonal distribution of high-latitude solar energy. Furthermore, the effect of the atmosphere is indeed important: at higher latitudes, significantly more solar energy is absorbed as the rays follow an oblique path through the atmosphere and are scattered or absorbed by an increased number of molecules.

Even so, the presence or absence of clouds in average local weather patterns can be much more important than either of the latitude effects: the maximum monthly average solar radiation received on a horizontal surface on the ground in eastern Washington State at 46 degrees latitude is considerably greater than that at Benin City, Nigeria, at 6½ degrees, and even the annual average insolation is greater in the more northern location.[9]

Maps can be drawn that represent the effects of average climate conditions, especially cloud cover, on the receipt of solar radiation in different portions of the globe. Solar fluxes at different seasons of the year are shown in figure 7-1. The maps show the generally large insolation of the Saharan and Sahel regions, even in winter. However, the specific energy end-uses to be addressed would need to be examined in order to choose especially likely areas for solar applications. For example, crop drying needs are more likely to be met economically by solar energy than water pumping; but they require sun at specific times of the year. Furthermore, the relatively low insolation indicated for parts of West Africa and Indonesia, compared to neighboring regions, shows that the tropics are not a uniformly desirable region for solar

[9] Tilting a solar collector upward or using sun-tracking collectors usually does not change most average results for the lower latitude regions, although these options may be very useful in higher latitude regions for increasing the insolation greatly in the low-sun periods when space heating is most needed. Therefore, for most developing areas, the use of fixed flat-plate collectors can well be the best choice; in cloudy areas, because of the great scattering of the solar radiation, fixed collectors may be superior even to the more expensive concentrating collectors, which tend to collect only direct solar radiation (Hans H. Landsberg, and coauthors, *Energy: The Next Twenty Years*, a study sponsored by the Ford Foundation and administered by Resources for the Future [Cambridge, Mass., Ballinger, 1979] p. 485).

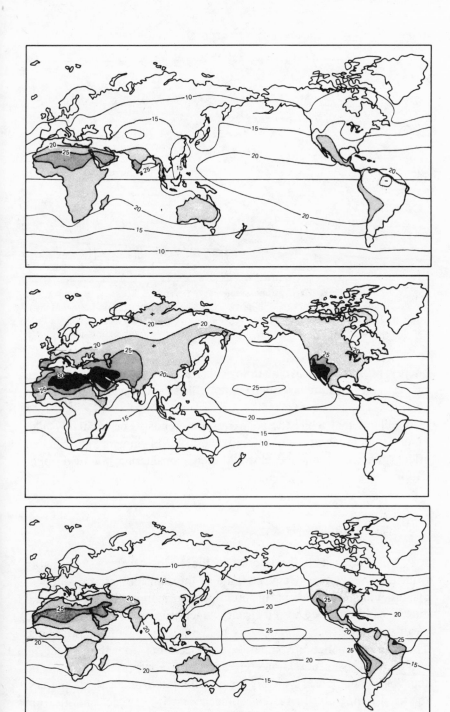

(Figure 7-1 concluded with caption on page 166)

165

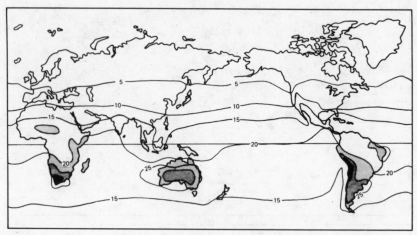

Figure 7-1. Solar Flux, March, June, September, December. From A. B. and M. P. Meinel, *Applied Solar Energy* (Philippines, Addison-Wesley, 1976). Meinel/Meinel, APPLIED SOLAR ENERGY: In Introduction, copyright 1976, Addison-Wesley Publishing Company, Inc., charts on pages 53, 54, 55. Reprinted with permission.

energy. Moreover, local climatological effects, too small to be displayed on the maps, result in large departures from the averaged-out insolation shown.

It follows that while the solar resource tends to be relatively high in quality in much of the developing world, local climatic and seasonal effects must be examined closely before considering practical applications.

Geothermal Energy

It is well known that the interior of the earth is warm—presumably from the radioactive decay of radium, thorium, potassium isotopes, and other unstable elements in natural rock formations. This geothermal energy can be tapped for electricity generation and other uses on the earth's surface, but utilizing this resource is not without problems. The average change in temperature, descending from the surface, is about 10° Celsius per kilometer, and since appropriate drilling and recovery technologies typically are now limited to about 10 kilometers, the total temperature rise that could be used to extract energy from "normal temperature gradients" is not high.

Practical use of geothermal energy at present, therefore, depends on finding hot spots in the earth's crust. Furthermore, no commercially feasible means has yet been found—although experiments on hot dry rock are now underway—to make use of geothermal energy, even at hotter-than-usual locations, unless steam or very hot water are naturally present. Indeed, all successful plants now operating—a handful in Italy, New Zealand, Japan, California, and Mexico—use sources of dry or wet steam from surface or near-surface wells.[10] Such sources are high in temperature and easy to use directly to generate electricity, but are relatively rare. This means that the greatest possibilities for the extensive use of geothermal technology lie in the hoped-for success of current pilot operations that use the much more common moderate-temperature waters or brines that are present in many areas, and also in geothermal sources lying at greater depths. Experiments are now underway in the United States and elsewhere.

Geothermal resources tend to be found in tectonically active regions of the world. (This could be taken as an instance of natural justice since those seismically active areas may be more dangerous for the use of nuclear power). Figure 7-2 shows "lines of fire" for bands of greatest seismic activity, together with a rough indication of the seismic data of some other areas of the world. The present nature of knowledge about geothermal engineering is limited, but one would expect promising geothermal resources most probably to lie within the seismic and semi-stable areas shown in the figure.

For most developing areas, the most likely use of geothermal is in hot steam or hot water sources that can produce electricity. Geothermal energy is widely used in Iceland, however, for space heating, and even to some extent in scattered locations in the United States and elsewhere.[11]

One uncertainty about geothermal energy is whether or not it is really "renewable." Natural geysers are probably replenished by the flow of rainwater percolating through the earth. Because of subsidence and pollution problems, it is becoming standard practice in modern geothermal installations to reinject the geothermal fluid after use. Whether this recharge mechanism or natural process will maintain a given geothermal source for an indefinite period (given the vast reservoir of heat that exists inside the earth) has not yet been determined.

[10] Richard Dorf, *Energy, Resources and Policy* (Reading, Mass., Addison-Wesley, 1978) p. 294.

[11] Schurr and coauthors, *Energy in America's Future*, chapter 11.

Figure 7-2. Geothermal Energy: Seismically Active Regions. Adapted from map on page 146 of Melville Bell Grosvenor, ed., *National Geographic Atlas of the World* (Washington, D.C., National Geographic Society, 1963).

LEGEND

Stable areas
Semi-stable areas
Seismic areas
Bands of greatest seismic activity

One kind of hybrid resource, geopressured gas, combines aspects of ordinary geothermal energy with natural gas production. Vast amounts of these reserves are thought to exist in the Gulf of Mexico coastal region of the United States. Formed by the weight of many layers of sedimentary strata surmounting reservoirs of water and methane, these fields could prove a valuable resource in other as yet unexplored areas. But exploration and development in developing areas should probably await the results of U.S. experiments.

For planning purposes, developing countries with obvious geothermal potentials will undoubtedly already have this option in mind. The not so obvious potential from second generation geothermal sources, such as lower temperature liquids, the natural gradients mentioned above, or dry rock or intrusions of magma, should be tracked closely by planners as experiments in the industrial countries proceed. The development of any of these resources could lead to the possibility of utilizing large amounts of the approximately 40 btoe of energy (at temperatures over 100°C) estimated to lie within 3 kilometers of the earth's surface.[12]

Biomass

Biomass resources include first of all the vast potential harvest from the world's forests and the residues from field crops and animal husbandry. These resources are going through a time of great change. While deforestation to extend agriculture and urbanization decreases forest inventories, a growing movement toward forest plantations gives promise of greatly increasing yields of wood for timber and pulp, and also for energy purposes. Second, the possibility of using special field crops for fuel production is being explored in many countries—in particular, the growing of sugar or starch crops to produce sugar for fermentation for fuel alcohol. The resource question can be considered separately for forests, residues (or wastes), and energy crops.

Forests

The first thing to be emphasized about world biomass resources is that data are very uncertain; the state of knowledge is far from ideal

[12] Electric Power Research Institute, "Unconventional Energy Resources." Prepared for the Conservation Commission World Energy Conference, Palo Alto, California, August 1977, p. 69.

Table 7-3. Estimated Existing Forest Biomass Resources by Continent

Continent	1973 total forest area (million hectares)	Theoretical annual increment (million metric tons oil equiv.)	1974 wood usage, fuelwood (million metric tons oil equiv.)	1974 wood usage, other (million metric tons oil equiv.)	1976 commercial energy (million metric tons oil equiv.)
Latin America	800	190–1,900	57	13	218
Africa	795	190–1,900	62	7	43
Asia (excl. Oceania)	506	120–1,200	147	24	251

Sources: Total forest area: Reidar Persson, *World Forest Resources* (Stockholm, Royal College of Forestry, 1974) p. 222; theoretical annual increment: see footnote 14 to the text; 1974 fuelwood and "other" wood usage derived from United Nations, Food and Agriculture Organization, *The State of Food and Agriculture* (Rome, 1976) p. 90.

in the industrial countries and is extremely poor for many developing areas. Data on total area in forests are regularly gathered by international agencies from national reports and other surveys, but the accuracy of many such reports is poor. Moreover, very little information is available on a consistent basis for key factors such as the annual increment of forest growth, much less—as would be theoretically most desirable—for the amount of wood that could be harvested on a sustained yield basis.[13]

Despite these formidable problems, it is useful to see what conclusions can be drawn from the available information. Table 7-3 shows data for three continents, including most market-economy developing areas (plus South Africa and Communist Asia). A theoretical calculation of the amount of energy that could be derived from the wood actually produced (the "annual increment") in such forest areas is shown. This includes a large allowance for error.[14] This estimate only shows the amount of wood that could be used in theory. In practice, many of the wood resources are in use already, would be too uneconomical to use, or could be lost through future deforestation.

The theoretical annual increments for the continents shown could provide an impressively large amount of energy, at least at the upper end of the range. The table shows estimates for recent use of forests for fuelwood and other uses (such as timber and pulp)—and then, for comparison, an estimate of the commercial energy demands for each of those continents. Even at the lower limits, sizable fractions of 1974 energy and wood products needs could be supplied in Latin America,

[13] For example, in addition to simple errors of estimation, data may include areas that once were forest and are no longer or areas that are merely planned for future forest growth. In addition the definitions and types of forests that are included may vary from tropical rain forests to savannas, with quite different consequences for biomass estimates. The total volume of living matter in the forests should be a prime datum, but has actually been carefully estimated only for some countries, mostly industrialized, and for scattered small areas elsewhere.

[14] In terms of volumes of wood produced, this theoretical increment is taken to be between one and ten cubic meters per hectare per year; these values are chosen on the basis of values reported to the FAO on temperate forests, which have approximately that range (Reidar Persson, *World Forest Resources* [Stockholm, Royal College of Forestry, 1974] pp. 204–205) and reports of average indigenous forest yields in tropical forests of 3–8 cubic meters per hectare per year (Norman E. Johnson, "Biological Opportunities and Risks Associated with Fast-Growing Plantations in the Tropics," *Journal of Forestry* vol. 74, no. 4 [April, 1976]). The volumes of wood are then converted to energy units by assuming that one cubic meter yields 0.232 toe. (Ten [9.67] gigajoules [0.232 metric tons oil equivalent] per cubic meter is from U.N. Food and Agriculture Organization, "Energy and Agriculture" in FAO, *The State of Food and Agriculture* (Rome, 1976) p. 2).

while all of Africa's needs would be suppliable—at least in a very theoretical sense. For Asia, the prospects are by no means derisory but are not quite as favorable.[15] Table 7-4 shows a similar calculation for the individual developing nations, looking at the theoretical potential from forests.[16] This table also includes potential contributions from animal manure and crop residues (see below).

Table 7-5 shows the theoretical biomass potential for selected developing countries plus the percentage of recently reported total energy demand that could theoretically be satisfied by these biomass resources.[17] Even if values near the upper end of the estimates are valid, the supply of forest (and crop and pasture residue) biomass does not, however, constitute a panacea for the energy crisis in the developing areas. Practical constraints are of immense importance; because of lack of data, no account at all has been taken of the cost of biomass fuel, which depends greatly on costs of harvesting and especially on the cost of transport. It is impossible to say from the sparse data available what typical fuelwood market (or opportunity) costs would be; estimates of $15 to $80 per metric ton oil equivalent for fuelwood sold in various developing areas have been quoted,[18] but local variations are obviously quite large. Furthermore, in examining these data, one must remember that if wood is to be a *renewable* resource, the forests of the developing world must be managed—rather than mined. Only such a truly renewable use can also prevent the exacerbation of the already great problem of deforestation, with all its attendant ills. This means that the strongest effort must be made to produce forest products on a sustained yield basis. Also, since little is known about managing primary forests, it probably means that in practice the greatest management efforts should be made with secondary forests.

Whatever the yields per hectare actually are, the land inventory is the critical constraint in the energy equation.[19] It is therefore of great

---

[15] Incidentally, if the wood usage figure for Asia (7.4 × 10⁸ cubic meters) is correct, it implies that the lower limit used here for the annual increment is definitely on the conservative side.

[16] Continent totals differ from table 2-5 somewhat because of differing definitions of continents and different years.

[17] Calculated on a gross energy equivalent basis.

[18] Elizabeth Cecelski, Joy Dunkerley, and William Ramsay, *Household Energy and the Poor in the Third World* (Washington, D.C., Resources for the Future, 1979) p. 52.

[19] The concept of "land" in a productive sense is of course not static. Many presently submarginal lands could be used if appropriate technology can be applied or developed.

Table 7-4. Potential Annual Energy Resources From Existing Forest and Wastes Biomass in Developing Countries

Continent, region, country	Total forest area (million hectares)	Annual increment[a] (million cubic meters)	Potential energy from forest growth[b] (10^{15} J)	Potential energy from animal manure (10^{15} J)	Potential energy from crop residues (10^{15} J)	Total potential annual energy from biomass[c] (10^{15} J)	Estimated 1976–77 fuelwood and charcoal consumption (10^{15} J)
Total Latin America	800.0	800–8,000	8,000–80,000	5,476.0	2,802.0	16,000–88,000	2,500
Central America							
Belize	2.0	2–20	20–200	N.A.	N.A.	20–200	N.A.
Costa Rica	2.2	2–20	20–200	31.6	16.4	68–250	24
Cuba	1.1	2–20	20–200	105.0	328.0	450–630	16
Dominican Republic	1.1	1–10	10–100	39.6	61.6	110–200	19
El Salvador	1.0	1–10	10–100	20.8	22.4	53–140	35
Guatemala	6.5	6–60	60–600	41.2	41.3	140–680	56
Haiti	0.2	0–2	2–20	30.6	22.7	55–70	42
Honduras	7.0	7–70	70–700	34.2	12.0	120–750	33
Jamaica	0.5	1–5	10–50	7.2	22.8	40–80	0.02
Mexico	40.0	40–400	400–4,000	645.0	405.0	1,500–5,100	90
Nicaragua	6.4	6–60	60–600	45.5	17.4	120–660	24
Panama	4.1	4–40	40–400	24.4	14.1	80–440	15
Puerto Rico	0.2	0–2	2–20	10.6	19.1	32–50	N.A.
South America							
Argentina	60.3	60–600	600–6,000	1,128.0	414.0	2,100–7,500	88
Bolivia	47.3	47–470	470–4,700	82.7	26.7	580–4,800	39
Brazil	320.0	320–3,200	3,200–32,000	1,950.0	948.0	6,100–35,000	1,600
Chile	5.0	5–50	50–500	86.2	37.7	170–620	33
Colombia	78.0	78–780	780–7,800	408.0	163.0	1,400–8,400	220
Ecuador	18.1	18–180	180–1,800	67.8	22.4	280–1,900	21

(continued)

Table 7-4. (continued)

Continent, region, country	Total forest area (million hectares)	Annual increment[a] (million cubic meters)	Potential energy from forest growth[b] (10^15 J)	Potential energy from animal manure (10^15 J)	Potential energy from crop residues (10^15 J)	Total potential annual energy from biomass[c] (10^15 J)	Estimated 1976–77 fuelwood and charcoal consumption (10^15 J)
French Guinea	8.6	9–90	90–900	N.A.	N.A.	90–1,900	N.A.
Guyana	18.2	18–180	180–1,800	6.2	28.7	210–1,800	0.2
Paraguay	21.0	21–210	210–2,100	88.5	16.6	310–2,200	33
Peru	87.0	87–870	870–8,700	142.0	85.9	1,100–8,900	63
Surinam	14.8	15–150	150–1,500	0.8	4.3	160–1,500	N.A.
Uruguay	0.5	1–5	10–50	232.0	25.5	270–310	10
Venezuela	48.0	48–480	480–4,800	248.0	46.4	1,000–5,400	80
Total Africa	795.0	795–7,950	7,950–79,500	2,963.0	895.0	12,000–83,000	3,100
Algeria	2.4	2–20	20–200	57.7	34.2	110–290	14
Angola	72.7	73–730	730–7,300	52.7	12.0	800–290	73
Benin	8.6	9–90	90–900	18.8	6.2	120–930	26
Botswana	11.0	11–110	110–1,100	41.4	0.1	150–1,100	8
Burundi	0.3	0–3	3–30	0.5	15.8	30–45	10
Cameroon	30.0	30–300	300–3,000	55.2	9.9	370–3,100	77
Central African Republic	28.0	28–280	280–2,800	9.6	3.1	290–2,800	22
Chad	16.5	16–160	160–1,600	69.1	1.1	270–1,700	38
Congo	27.0	27–270	270–2,700	1.1	4.1	280–2,700	20
Djibouti	—	—	—	N.A.	N.A.	N.A.	N.A.
Egypt	—	—	—	98.3	182.0	280	1
Equatorial Guinea	1.0	1–10	10–100	0.2	0.2	10–100	N.A.
Ethiopia	33.0	33–330	330–3,300	586.0	67.1	980–4,000	250
Gabon	24.8	25–250	250–2,500	0.5	0.4	250–2,500	12

Gambia	0.1	0–1	1–10	5.4	0.4	7–16	3
Ghana	12.0	12–120	120–1,200	29.9	17.1	170–1,200	120
Guinea	17.1	17–170	170–1,700	26.6	11.3	210–1,700	29
Guinea–Bissau	1.1	1–10	10–100	5.7	0.8	17–110	5
Ivory Coast	19.0	19–190	190–1,900	14.8	16.5	220–1,900	53
Kenya	1.9	2–20	20–200	149.0	37.6	200–400	130
Lesotho	—	—	—	16.9	2.6	20	N.A.
Liberia	2.5	2–25	25–250	2.1	4.1	31–260	15
Libya	0.5	0–5	5–50	15.5	10.3	31–75	4
Madagascar	12.5	12–120	120–1,200	167.0	46.7	330–1,400	55
Malawi	7.0	7–70	70–700	13.0	15.3	98–730	33
Mali	4.5	4–40	40–400	89.3	3.2	110–490	30
Mauritania	—	—	—	51.9	0.1	52	6
Morocco	5.2	5–50	50–500	27.0	5.3	82–530	30
Mozambique	66.5	66–660	660–6,600	26.5	22.3	710–6,600	90
Namibia	10.0	10–100	100–1,000	N.A.	N.A.	100–1,000	N.A.
Niger	4.0	4–40	40–400	66.5	1.8	110–470	25
Nigeria	34.4	34–340	340–3,400	277.0	79.2	700–3,800	700
Rwanda	0.5	0–5	5–50	14.1	3.3	22–67	42
Sénégal	5.4	5–50	50–500	53.0	3.4	110–560	25
Sierra Leone	0.3	0–3	3–30	5.7	9.1	18–45	27
Somalia	0.2	0–2	2–20	98.9	3.5	100–120	35
Sudan	42.0	42–420	420–4,200	327.0	20.5	770–4,500	230
Swaziland	—	—	—	11.2	10.8	22	5
Tanzania	38.9	39–390	390–3,900	217.0	96.1	700–4,200	400
Togo	3.6	4–40	40–400	8.8	4.8	53–450	10
Tunisia	0.3	0–3	3–30	30.5	41.6	75–100	19
Uganda	1.9	2–20	20–200	76.5	15.6	110–290	0.50
Upper Volta	3.5	3–30	35–350	39.3	1.6	76–390	43
Zaire	180.0	180–1,800	1,800–18,000	29.6	31.9	1,900–18,000	0.30
Zambia	37.3	37–370	370–3,700	1.8	9.6	380–3,700	40
Zimbabwe	27.8	28–280	280–2,800	73.8	32.1	390–2,900	60

(continued)

Table 7-4. (continued)

Continent, region, country	Total forest area (million hectares)	Annual increment[a] (million cubic meters)	Potential energy from forest growth[b] (10^15 J)	Potential energy from animal manure (10^15 J)	Potential energy from crop residues (10^15 J)	Total potential annual energy from biomass[c] (10^15 J)	Estimated 1976-77 fuelwood and charcoal consumption (10^15 J)
Total Asia	506.0	506–5,060	5,060–50,600	11,018.0	11,925.0	28,000–74,000	5,600
Near East (Asian)							
Afghanistan	0.7	0–7	7–70	124.0	44.1	180–250	64.0
Bahrain	—	—	—	0.1	N.A.	—	N.A.
Iran	4.0	4–40	40–400	261.0	276.0	580–940	22.0
Iraq	1.5	1–10	15–150	98.4	48.7	160–300	0.1
Jordan	—	—	—	4.7	2.5	7	0.03
Kuwait	—	—	—	1.1	—	1	N.A.
Lebanon	0.1	0–1	1–10	3.7	3.0	8–17	0.7
Oman	0.1	0–1	1–10	2.3	0.1	3–12	N.A.
Qatar	—	—	—	0.4	N.A.	—	
Saudi Arabia	1.2	1–10	12–120	24.6	6.5	43–150	N.A.
Syria	0.5	0–5	5–50	11.5	74.4	91–140	0.5
Turkey	18.2	18–180	180–1,800	417.0	679.0	1,300–2,900	N.A.
U.A.E.	—	—	—	N.A.	N.A.	N.A.	N.A.
Yemen (AR)	—	—	—	52.5	12.9	65	N.A.
Yemen (P.D.R.)	2.6	3–30	26–260	5.4	0.6	32–300	N.A.
East Asia							
Bangladesh	2.3	2–23	23–230	492.0	369.0	880–1,100	150
Bhutan	3.0	3–30	30–300	4.0	7.6	42–310	N.A.
Brunei	0.4	0–4	4–40	0.5	0.1	5–41	N.A.
Burma	45.0	45–450	450–4,500	173.0	176.0	800–4,800	210

176

China	80.0	80–800	3,350.0	800–8,000	5,096.0	9,200–16,000	1,500
India	75.4	75–750	4,250.0	750–7,500	3,043.0	8,000–15,000	1,300
Indonesia	124.7	125–1,250	219.0	1,250–12,500	559.0	2,000–13,000	1,200
Khmer Rep.	13.2	13–130	47.2	130–1,300	N.A.	180–1,300	44
Korea N.	9.0	9–90	23.3	90–900	108.0	220–1,000	50
Korea S.	6.7	7–70	40.1	70–700	206.0	320–950	80
Laos	14.8	15–150	33.1	150–1,500	16.2	200–1,500	33
Malaysia	23.6	24–240	20.7	240–2,400	40.2	300–2,500	61
Mongolia	15.0	15–150	113.0	150–1,500	10.9	280–1,500	15
Nepal	4.7	5–50	184.0	50–500	60.1	290–740	96
Pakistan	2.3	2–20	494.0	20–200	450.0	960–1,100	93
Philippines	15.9	16–160	177.0	160–1,600	289.0	630–2,100	250
Sri Lanka	2.5	2–20	41.9	20–200	24.3	86–270	46
Taiwan	2.0	2–20	N.A.	20–200	N.A.	20–200	N.A.
Thailand	29.0	29–290	205.0	290–2,900	308.0	800–3,400	180
Vietnam	7.5	7–70	143.0	70–700	13.5	230–860	180

Notes: This table gives the potential energy that theoretically might be extracted from annual forest growth and vegetable and animal wastes in developing nations. The figures do not represent realistic estimates of future production, because constraints involving alternate economic uses, harvesting and transport feasibility, and potential end-use devices are not taken into account. To convert to million metric tons oil equivalent, multiply the amount in units of 10^{15} joules (J) by 0.0232; for example, for Belize, the total potential energy from biomass (next to last column) is 464,000 to 4,640,000 tons oil equivalent (0.46–4.6 mtoe). N.A. = not available. Dashes = not applicable.

Sources: Total forest area from Reidar Persson, *World Forest Resources* (Stockholm, Royal College of Forestry, 1974) table 5; it includes "closed forests" and "open woodlands." Energy from manure and crop residues from T. B. Taylor Associates in a report to The Office of Technology Assessment, U.S. Congress, "Survey of Biomass Energy Programs and Use in the Developing Countries," December 23, 1978, review draft, pp. 23–37. Consumption from World Bank, *Prospects for Traditional and Non-Conventional Energy Sources in Developing Countries,* Staff Working Paper No. 346 (Washington, D.C.) p. 37.

[a] Volume of timber assessed to be capable of increase at 1–10 cubic meters per hectare.

[b] Assumes an energy content of 10 gigajoules per cubic meter.

[c] Energy from theoretical incremental growth of forests plus energy from animal wastes and crop residues.

Table 7-5. Theoretical Contribution of Existing Forest and Wastes Biomass to Total Commercial and Noncommercial Energy Demand in Selected Countries, 1977

Country	Theoretical energy supply from biomass (million metric tons oil equivalent)	Theoretical percentage of total energy consumption satisfiable by biomass
Algeria	2.6–6.7	12–31
Argentina	50–170	120–390
Brazil	140–800	120–710
Colombia	30–200	150–920
Egypt	6.5	46
India	210–370	160–270
Indonesia	50–300	69–410
Iran	13–22	27–44
Jamaica	0.9–1.9	31–63
Kenya	4.6–9.3	100–210
Korea, S.	7.4–22	21–63
Mexico	35–120	43–150
Nigeria	16–90	67–360
Saudi Arabia	1–3.5	6–19
Thailand	19–80	150–590
Turkey	30–70	92–200
Venezuela	23–130	74–400

Sources: Table 7-4 in this chapter and data derived passim from the following documents: International Energy Agency/Organisation for Economic Co-operation and Development, *Workshop on Energy Data of Developing Countries* (Paris, OECD, 1979); International Energy Agency/ Organisation for Economic Co-operation and Development, *Energy Balances of OECD Countries 1974–1978* (Paris, OECD, 1980).

importance to know how production of fuelwood, for example, could be increased by improving forest management practices. There is evidence to show that the yields posited above of 1 to 10 cubic meters (0.2 to 2.0 metric tons oil equivalent [toe]) per hectare are small compared with what improved cultivation might yield. Even without special efforts to improve yields, the statistics could probably be revised upward by counting in the small-diameter wood and dead wood that is often not taken into account in timber-oriented statistics. Such wood is now utilized as fuel by the rural population of developing areas and can always be used in energy schemes. Similarly, if the canopies, twigs, and roots were used, additional energy could be obtained. An analysis of actual existing forest area in the United States found that such measures using "junk wood" could increase the present overall average yield from 1.2 tons oil equivalent per hectare to 2.8.[20] Other improvements, such as fertilization, could raise that to 5 toe, while genetic measures might raise the yield to 7 toe per hectare.[21]

In a similar fashion, some specialized tree crops in the tropics have been estimated to produce high yields, even without great care in cultivation: *Pinus caribaea,* 4 toe per hectare; *Gmelina arborea,* 6.5 toe; and *Albizia falcataria,* 8 toe.[22]

It should be noted that long-term yields from plantations, especially in the context of highly leached tropical soils, may not approach these theoretical levels. The long-term effects of depletion of nutrients and the dangers of diseases and pests in attacking a sylvan monoculture remain for the most part to be assessed.[23] Nonetheless, it has been estimated that in the coming decades, plantations in those Asian countries with market economies will come to supply 25 percent of the total annual increment in forest products.[24] Such projections give hope that the forest biomass resource can be more effectively used to provide energy as well as timber, pulp, and other commercial forest products. These estimates, however, hide a disturbing problem: deforestation in developing areas in Asia is projected to cancel out any increases in the annual increment from better cultivation or harvesting.[25] Since much of this deforestation will presumably be carried out in order to extend agriculture, the complex problem of food versus fuel makes its first appearance in this study: it will be discussed below and in chapter 8.

Residues (Wastes)

For crop and animal residues, also shown in table 7-4 and included in the totals in table 7-5, the situation is somewhat different. It must first be emphasized that the single-figure estimates for the amount of

[20] William F. Hyde and Frederick J. Wells, "The Potential Energy Productivity of U.S. Forests," mimeo (Washington, D.C., Resources for the Future, 1975) p. 8.

[21] Forest "wastes" of course return inorganic nutrients to the soil when they decay. These would have to be replaced if the wood is removed to utilize the organic part for energy purposes if balances are to be maintained. This question of recycling is now receiving close scientific study (Hyde and Wells, "Potential Energy Productivity," p. 18).

[22] Johnson, "Biological Opportunities."

[23] Ibid.

[24] United Nations Food and Agriculture Organization, *Development and Forest Resources in the Asian and Far East Region* (Rome, 1976) p. 16.

[25] Ibid.

animal and crop residues shown in table 7-4 are not thereby more exact than the range shown for wood estimates; in fact, they may be even less exact, because the wastes are even less likely to enter commercial markets. These residues, it must be stressed, are *potentially* available for energy use in a much more conditional way than forest products. While local situations vary, underutilized forests containing tons of decaying dead growth will continue to be uneconomic and therefore unavailable in a practical sense even in energy-deficient countries, as long as the forests are sufficiently far from population centers. Crop residues, on the other hand, may be unavailable for new energy uses for the opposite reason: they are generally close at hand to the farmer and are therefore often fully utilized for fertilizer, animal feed, or in many countries, as a replacement for locally scarce firewood or—depending on tastes—as a more desirable fuel for cooking fires. In South Asia, for example, much use is made of animal and crop residues as fuel, in contrast with Latin America. In either case, the potential wastes shown as "available" are quite likely already being put to some economic use, so that diversion to new or expanded energy uses is not free of opportunity cost and may not be practicable in real-life cases.

Energy Crops

Arable land could be used to grow field crops designed to be harvested for energy use instead of for food and fiber. The difficulty is that the value of most biomass used as fuel is only $15 to $80 per metric ton oil equivalent.[26] Measured in energy units, the values of most field crops, however, are far greater than this in the existing commercial markets for food and fiber. This means that field crops as energy sources must have some special characteristic over and above the thermal energy that they can produce. One such characteristic would be their ability to grow where no other crop will. Experiments are now being carried out on water crops, such as water hyacinth, kelp, and various algae.[27]

An especially valuable type of energy crop would be one that produces hydrocarbons directly, and therefore promises to be a

[26] Cecelski and coauthors, *Household Energy*, p. 52.
[27] See, for example, Melvin Calvin, "Petroleum Plantations for Fuel and Materials," *Bio-Science* vol. 29 (Sept. 1979) pp. 533–38 for a description of hydrocarbon-containing and other energy crops.

replacement for liquid fuels like gasoline, which is valued at about $350 per toe. We do not consider this option in detail here, even though experiments with species such as jojoba and euphorbia will be followed with interest by energy planners.

Finally, there are food crops that, while not producing hydrocarbons directly, produce carbohydrates in the form of ethanol and grain alcohol that can be readily turned into a prime substitute for liquid transportation fuels. Such sources of alcohols can be sugar crops like sugar cane, or starchy crops like food grains and tubers. Such crops often have high values in conventional food or fiber uses—the energy in sugar at 10¢ per pound costs about $400 per ton oil equivalent—and so diverting them to energy tends to be relatively expensive. But sugar and grain alcohol technology is well established, and such crops could play an important role in self-sufficiency schemes for liquid fuels. This option is discussed in more detail in chapter 8, where the economic problems of tradeoffs between food and fuel are reviewed and the potential size of this type of biomass resource is sketched.

Conclusions

The renewable energy resource picture is of critical importance for energy policy decisions in developing areas. But the status of the resource endowments considered in this chapter is varied. The hydraulic resource is an old established one. Developing areas with hydroelectric resources are aware in a general way of existing potential. But one thing that has changed is the economics: oil-fired generation as an alternative is less and less attractive. The potential hydroelectric resource under changing energy economics appears very large in relation to foreseeable electricity needs in many countries. This calls for conventional hydroelectric sites to be reexamined and the intriguing new mini-hydro option to be given serious study.

Geothermal power is a technology that can be very efficient, given the right type of geothermal resources. Developing areas can safely invest in new generating facilities based on geothermal steam. However, if lower grade hot water sources become established as commercially feasible, geothermal power could make a very great contribution to electricity needs in many developing countries. Still, finding liquid-dominated geothermal resources will require extensive exploration efforts and some proof of engineering success in current pilot operations.

In the tropical areas of the developing world the solar resource suffers less from seasonal variations than it does in the temperate world. However, the effects of cloudiness must be taken into account in many tropical locations, and the need for specialized end-uses like crop drying must of course take into account correlations with wet and dry seasons. The size of the resource is not doubted: questions of technology are paramount (see chapter 8).

The forest resource and pasture and crop residue biomass resource are potentially large. However, effective transport and utilization costs will often be prohibitive; a close study of the economics of forest use remains one of the critical needs for energy planning in many developing countries.

Much improvement in terms of wood yield per land unit can be achieved by plantation schemes, although the full story on tropical tree plantations has yet to be written. For many countries, field crops could also provide sugar or starch for liquid-fuel replacements for gasoline and other liquid fuels.

It follows that the energy planners must generally deal with substantive problems of gathering information on the basic potential of renewable energy resources. They must also face complex issues involving the analysis of the practical costs of these types of energy supplies in the context of new technologies and competing uses for land and other essential resources.

8

Supply and Costs of
Key Energy Technologies

The resources, mineral and renewable, described in chapters 6 and 7, cannot be used directly in energy end-uses. In all cases, there is some type of technology involved, even if the "technology" consists of merely a few stones to form a campfire. In some cases, the energy is typically processed into other more complex forms before the final usage.

Many energy-processing steps, such as petroleum refining, are very complex activities that play an important role of their own in the economic life of a country. Other forms of technology vary from intermediate steps such as roadside charcoal making to a wide spectrum of end-uses such as furnaces and diesel engines.

Our treatment of supply and cost questions is intended to be selective, rather than encyclopedic, and not all areas of technology are considered here. In particular, some popular new options, such as direct solar energy for space heating and water heating—which could be useful for hospitals, schools, and factories—are not discussed. Industrial cogeneration of electricity, while possibly a promising option in some developing countries, is not treated. Nor are possibilities for using pyrolytic oils and gases. But such technologies can still be recommended as important options in particular contexts in developing countries.[1] As in chapter 7, we have used as criteria for inclusion the potential importance of a technology either as a major energy source or as a key element in overall planning for social and economic development.

---

[1] See for example William Ramsay, "Prospects for Biomass Fuels in Developing Areas (I): Methanol and Other Thermochemical Conversion Products," Paper presented at the Ford Foundation meeting on Energy in Less Developed Countries, Washington, D.C., July 1979 (Washington, D.C., Resources for the Future).

Three key areas of technology involving a critical energy sector or a new approach to technology will be discussed here. The electricity sector is crucial to development of modern industry and will be examined in the light of past and expected oil price increases. Alcohol fuels represent a possible alternative way to fuel the road transport sector, which is one essential element of infrastructure required for economic growth. Finally, the general grouping of resources and appliances that are often called "appropriate village technologies"— or technologies peculiarly appropriate to rural use—will be examined as a possible culturally compatible answer to the supply of rural energy needs for both subsistence and development.

Electricity

Part of the story of the supply of electricity—an extremely versatile energy form—can be read in our earlier discussion of fossil fuels, uranium, hydroelectric potential, and other renewable sources. For purposes of energy policy decisions, however, the supply of electricity must take into account the total costs of the primary resource plus the cost of facilities for generation, transmission, and distribution. These include not only the financial cost, but also the many important health and environmental effects of generating electricity. This is especially true for generating electricity by coal and nuclear power, but also to some extent for other fossil fuels and hydropower and for some of the other renewable sources.

To review the general cost outlook for technologies for central-station electric generation as applied to the developing world in general, one must abstract from actual spot prices of commodities and services needed for capital construction in specific developing areas. Also, fuel and operating costs should be very sensitive to local and national conditions. But capital cost comparisons based on recent detailed analysis in the the United States—which are readily available—should be useful in analyzing the developing-area outlook. Even for U.S. plants, costing is complicated by treatment of inflation (especially in regard to interest rates), taxation, and utility accounting procedures. However, the costs shown in table 8-1 should give a good idea of the relative capital costs of plants of different sizes for various technologies.

How do we expect these costs to compare with capital costs in developing areas? Recent work by the World Bank estimates cost

Table 8-1. Estimated Mid-1980 Capital Costs of Power Plants, Based on U.S. Experience

Technology	Rating (megawatts of electricity)	Average U.S. cost ($1975 per kilowatt)
Oil[a]	225	340
(boiler)	400	290
	550	270
	1,100	230
Gas turbines[b]	90	160
Coal[c]	400	490 (+ 100 for scrubbers)
	550	450 (+ 90 for scrubbers)
	1,000	390 (+ 80 for scrubbers)
Hydroelectric[d]	various	500–600
Nuclear[e]	600	640
	1,100	550
Geothermal[f]	50	650
	100	530

[a] Center for Energy Studies, "Future Central Station Electric Power Generating Alternatives," (Austin, Tex., University of Texas for Resources for the Future, 1979) table 4.3-8.

[b] Ibid., sec. 4.3, pp. 4-203 and 4-206.

[c] Ibid., table 4.3-16.

[d] For a limited sample see Federal Power Commission, *Hydroelectric Plant Construction Cost and Annual Production Expenses,* Eighteenth Annual Supplement (Washington, GPO, 1976) p. XIII. Costs are very dependent on site and possible multiple uses. See text.

[e] Center for Energy Studies, "Future Central Station," table 4.1-7.

[f] Derived from Ibid., table 4.4-4. Note that costs differ considerably with fluid temperature, technology, and depth of resource.

"adders" over a European base ranging from 15 to 70 percent for several different developing areas.[2] But these apparent cost premiums could in fact be explained largely by differences in the type and size of power plants.[3] Therefore, the industrial-country (U.S.) costs shown in the table may still give a useful indication of likely cost relationships. Interest rates in local capital markets, local wage rates and labor availability, transportation costs, and import duties would of course have to be taken into account for any more precise national analysis.

[2] David Hughart, *Prospects for Traditional and Non-Conventional Energy Sources in Developing Countries,* Staff Working Paper No. 346 (Washington, D.C., World Bank, 1979) p. 77.

[3] For a scaling law exponent of 0.75, a 550-Mw plant costs 19 percent more per kilowatt than an 1,100-Mw plant. See Center for Energy Studies, "Future Central Station Electric Power Generating Alternatives" (Austin, Tex., University of Texas for Resources for the Future, 1979) table 4.3-8.

Total costs of generation include not only capital costs but also fuel (and fuel inventory storage) and operating and maintenance costs. For all fossil and uranium plants, fuel costs tend to be the dominant factor—often, as for oil, far exceeding capital costs. These fuel costs tend to vary widely between geographical areas—even within the United States,[4] and it is difficult to generalize about them for all developing countries. However, some discussion of total costs is given below—especially for "free-fuel" renewable energy sources.

From Oil and Gas

Oil-fired electricity is one of the mainstays of electricity production in many developing areas and as such is a potential hostage to future rises in the price of oil. Consequently, even the usually moderate capital costs for oil-fired plants (table 8-1 shows $290 per kilowatt [kW] for a 400-megawatt [Mw] plant) will often fail to make the option very attractive. Even in countries with ample domestic oil supplies, the oil should in most cases be valued at its high price in the international market, so the opportunity cost of using such oil for domestic electricity generation could be large.[5] Nevertheless, there may be no alternative to the continued use of this technology in many countries.

However, there are some bright spots on the horizon for the country that must use oil to generate its electricity. Combined cycle systems, using combustion gases first to drive the gas turbine and only afterward to heat a conventional boiler to drive a steam turbine, can achieve significant cost reductions by means of increasing efficiencies, especially for the usually expensive "intermediate-duty" generating cycle.[6]

The basic capital cost situation is very similar for generating plants fired by natural gas in that capital costs for ordinary gas steam boiler plants should be similar to those for oil. However, in some countries there are gas fields that cannot be fully utilized by existing distribution and conversion systems, and the opportunity cost of the gas fuel may in consequence be low. Furthermore, in that case gas turbines—which

---

[4] Ibid., tables 4.3-11, 4.3-18.

[5] The economics, however, may depend on the price of *residual* oil, not crude oil. Because of refinery location, specific gravity of crudes, or local product mix demands, local prices of residual could be significantly lower than those of crude or other products.

[6] If efficiencies climbed from 36 percent to 44 percent, for example, fuel costs would drop by 20 percent (Center for Energy Studies, "Future Central Station," table 4.3-3, figure 4.3-14).

are relatively fuel intensive—might be used instead of boilers, since their capital costs are quite low, as shown in table 8-1.

From Coal

Although coal and coal-like resources are presently of relatively little importance in most developing countries, it was emphasized in chapter 6 that new exploration efforts might turn up significantly larger resources and that imported coal could be a desirable option in many cases. The economics of coal for boiler use may then become of increased interest as time goes on. Costs are often dominated by the fact that coal is an inconvenient fuel to handle, and its rather modest energy content (relative to oil) per ton means that it will be relatively expensive to transport to many end-use applications. Therefore, the local price of coal may be relatively low in energy terms—reflecting the fact that its usual usage patterns may be dominated by a restricted set of energy users: heavy industries, rail transportation, and electrical utilities. Electricity planning can often take advantage of this price differential through building mine-mouth generating stations.

At any rate, the capital costs for coal-fired power plants, while higher than those for oil and gas, are somewhat less than those for nuclear stations. As shown in table 8-1, a typical small plant (of 400 megawatts) in the United States would cost about $490 per kilowatt, or, when it includes the type of air pollution equipment now used in the United States, about $590 per kilowatt. Operating and maintenance costs are probably somewhat larger for scrubber technologies than for some other fuels, but average expected values for *all* non-fuel costs over future decades for the United States have been projected at about 1.3 cents per kilowatt hour.[7] Adding in fuel costs by taking a rough average of American costs of coal of about $40 per metric ton oil equivalent (toe), this amounts to a total cost of 2.2 cents per kilowatt hour. Nonetheless, it should be noted that even in the relatively coal-rich United States, the fuel costs vary by about 50 percent among the nine main regions used for utility statistics. For developing areas, higher prices of imported or local coal can be taken into account by assuming that each additional $4 per ton oil equivalent (or $3 per ton of average coal) raises the electricity cost by about 0.1 cent for each

[7] Adapted from Center for Energy Studies, "Future Central Station," table 4.3-18, for 400 Mw taking a 10 percent (deflated) capital charge rate.

kilowatt hour (kWh). Of course, more modern methods of burning coal, such as fluidized-bed combustion, or the transformation of coal into other fuels by solvent refining or other synthetic fuel methods, may revolutionize the use of this mineral fuel. At this time, such techniques are not yet established and are premature for serious consideration by most developing-country planners.

From Hydroelectric Plants

The costs of hydroelectric plants vary enormously as a function of local geography and the nature of the water body involved, as pointed out in chapter 7 in evaluating the potential for hydraulic resources in the developing areas in general. In spite of that, some potentially useful generalizations can be made. Typical capital costs for a small sample of hydroelectric plants over a number of years in the United States ranged from $500 to $600 per kilowatt;[8] this range is roughly equivalent to that for a sample of eighteen older plants of various sizes, taking inflation into account.[9] These costs were for a whole variety of different sites, with different construction expenses—depending especially on terrain and in some cases on allocation of costs between multiple reservoirs or multiple uses of a single reservoir. Of this total, costs of "reservoirs, dams, and waterways," and "railroads, roads, and bridges" also vary as a proportion of the costs, but are typically from one-third to two-thirds of the total.[10]

To translate these capital costs into unit costs of electricity, one can assume, based on international statistics,[11] that the plant runs an average of about half the hours in the year. This means that at a capital charge of 10 percent per year, the unit cost of electricity from hydroelectric plants is about about 1.1 to 1.4 cents per kilowatt hour, or, including operating and maintenance costs, perhaps 1.2 to 1.5 cents per kilowatt hour.[12]

[8] Federal Power Commission, *Hydroelectric Plant Construction Cost and Annual Production Figures*, Eighteenth Annual Supplement (Washington, D.C., GPO, 1976), p. XII.

[9] Center for Energy Studies, "Future Central Station," table 4.6-2.

[10] Federal Power Commission, "Hydroelectric Plant," p. 1ff.

[11] World Energy Conference, *Survey of Energy Resources* (New York, United States National Committee of the World Energy Conference, 1974) p. 163.

[12] Federal Power Commission, *Hydroelectric Plant*, table 5, p. xv.

These estimates can rise if the hydroelectric sites are not favorable and dams are therefore more expensive to build. Furthermore, the lifetime (and therefore the average costs) of hydroelectric projects depends on rates of siltation and other local factors that may vary from place to place. Nevertheless, there are some general deductions that can be made for hydroelectric costs—neglecting for the moment environmental costs such as siltation damage. The most obvious advantage is the cost that does not exist: the zero cost of fuel. The second and not so obvious advantage is the cost of operation and maintenance, which has been given as about 0.08 cents per kilowatt hour for recent U.S. experience; this is lower than maintenance costs for coal and nuclear plants under current operating conditions, especially if scrubbers for coal plants are included.[13]

Finally, although capital costs can be high, the nonimpoundment capital costs (especially of generating equipment) should be expected to run from as low as 0.3 cents per kilowatt hour to at most about one cent per kilowatt hour, based on the proportions observed in these recent U.S. data.[14]

The difference between 0.3 and 1.0 cent per kilowatt hour for equipment and other nonimpoundment costs is important for evaluating prospects for new plants that would have been too expensive in the era of cheap oil. If the low end of the range of equipment and other nonimpoundment costs (0.3 cents) were to hold, one might speculate that economically feasible sites would be relatively easy to find.[15] Developing countries might also benefit if the lion's share of the costs go for the impoundments as most of the labor and materials for dam building are local and the extra costs of impoundment would not imply a proportionate increase in foreign exchange burdens.

Hydropower on a small scale can also be utilized as a "central station" source. However, economies of scale hold for hydropower as well as for other electricity technologies, and the most promising uses of mini-hydro facilities may be in localized systems. Therefore, we discuss the outlook for them below under "village technologies."

---

[13] Sam H. Schurr, Joel Darmstadter, Harry Perry, William Ramsay, and Milton Russell, *Energy in America's Future: The Choices Before Us* (Baltimore, Johns Hopkins University Press for Resources for the Future, 1979) pp. 275, 283.

[14] Federal Power Commission, "Hydroelectric Plant," p. 1ff.

[15] Since at zero impoundment costs no sites can be used, and at infinite costs there are a very large number of feasible sites, one might expect the "feasibility density" of sites to rise as impoundment costs rise, other things being equal.

From Nuclear Power

The outlook for the use of nuclear energy in developing areas can best be viewed as confused. On the one hand, although no longer holding out the prospect of power so cheap that it would not have to be metered, nuclear power still appears to be the cheapest source of electricity per kilowatt hour in the United States and probably in the other OECD countries[16]—except where it competes with inexpensive coal or low-cost hydro sites or with natural gas that does not have access to high-priced world markets. Table 8-1 shows that capital costs for a 600 megawatt nuclear plant are about $640 per kilowatt hour under U.S. conditions. Total unit financial costs of nuclear electricity are about 1.8 cents per kilowatt hour.[17]

At the same time, with the costs of construction in general, and of nuclear plants in particular, rising at rates higher than the general rate of inflation, the continued economic viablity of nuclear power depends on this trend not getting out of hand. Furthermore, the problems experienced by the industry regarding reliability of equipment operation—for example, cracking in the stainless steel pipes of boiling water reactors in recent years—raise questions about both costs and safety. Indeed, recent uncertainties about the functioning of both equipment and personnel in ensuring nuclear safety are the greatest single worry for nuclear energy planners: the Three Mile Island accident is only one event in a growing pattern of concern about safety matters.[18] And although some nuclear experts believe that the waste problem has been greatly overblown, the lack of effective policy decisions has succeeded in arousing serious public concern about long-term disposal.[19] At any rate, treating such problems is difficult enough in an industrialized

[16] See Sam Schurr and coauthors, *Energy in America's Future*, p. 288.

[17] From Center for Energy Studies, "Future Central Station," table 4.1-8, for 600 Mw, with a 10 percent capital charge adopted here. See also chapter 9 in Schurr and coauthors, *Energy in America's Future*, for a discussion of costing problems. This costing does not follow utility accounting conventions, but does provide an accurate opportunity cost for a usable comparison with new technologies.

[18] Schurr and coauthors, *Energy in America's Future*, pp. 353–360. Subsidies to nuclear power from government restrictions on liability limits are negligible—if reactors are relatively safe. For safe reactors, one expects few insurance payoffs. If reactors are not safe, on the other hand, higher liability limits may be irrelevant in the context of public perception of "acceptable risks."

[19] Waste disposal costs should be manageable (William Ramsay, *Unpaid Costs of Electrical Energy* [Baltimore, Johns Hopkins University Press for Resources for the Future, 1979] chapter 5).

context. Where trained cadres of technicians and engineers are in short supply, safety and reliability concerns can loom very large in planning decisions.

Even though the electricity needs of most developing countries would involve only a small number of nuclear reactors, developing-area planners must also take into account the international concerns that have arisen over the possible spread of nuclear power enrichment and reprocessing facilities.[20] With a view to discouraging nuclear proliferation, recent U.S. government policy has restricted exports of uranium, enriched uranium, and nuclear equipment and technology to countries that decline to renounce policies that many supplier nations consider potentially dangerous from the proliferation point of view. Such steps underscore the dependence on supplier-nation policy that falls to the lot of developing nations adopting the nuclear option. To be sure, restrictions by suppliers can be avoided by adherence to the Nuclear Nonproliferation Treaty or acceptance of equivalent "full-scope" international safeguards.[21] Furthermore, some aspects of exclusive dependence on U.S. supplies are fast being removed as the ability to enrich uranium and to reprocess nuclear fuel is spreading to newly constructed plants in Europe and Japan.

Nuclear power for a developing area therefore involves many types of costs that do not show up on the ordinary balance sheet, especially an entire array of health and environmental costs (see also chapter 9). Furthermore, the ability of nuclear power to generate electricity at about 2 cents per kilowatt hour in the United States cannot necessarily be transferred easily to developing economies. For example, costs of construction and operation, including the training or recruitment of qualified engineers and other personnel, could be substantially larger.

The question of size is also important, since there is a considerable economy of scale in nuclear plant operation. Only a few countries would have national systems with capacity as large as 5 gigawatts (GW) in the next few decades, and 10 percent reserve reliability requirements would restrict individual plant size to 500 megawatts or thereabouts. Therefore many developing-area grids would have to use reactors much smaller than the 1,200 megawatt types now often standard in the United States. However, this size factor should not be

[20] Ibid, chapter 6.
[21] See, for example, Ford Foundation, *Nuclear Power Issues and Choices,* report of the Nuclear Energy Policy Study Group, administered by MITRE Corporation (Cambridge, Mass., Ballinger, 1977) chapter 9.

overemphasized because a large proportion of the reactors now generating power in the world are below 500 megawatts in capacity or even below 200 megawatts. These smaller reactors are indeed not as economical as the larger ones: table 8-1 shows the 600 megawatt reactor as 16 percent more expensive (per kilowatt of capacity) than the 1,100 megawatt plant, and a 250 megawatt reactor can be expected to cost 50 percent or so more per kilowatt than a 1,000 megawatt reactor. Moreover, even though nuclear systems manufacturers are currently emphasizing larger designs, the resulting costs—even up to $1,000 per kilowatt ($1975, with deflated interest rates)—might not be prohibitive in some areas, and smaller reactors could still conceivably be more economic than some alternative methods of power generation.

A final mundane factor that might be considered in any policy decision is the generally depressed nature of the nuclear vendor industry at the present time. When domestic sales are slow, reactor manufacturers depend more and more on foreign purchasers. As many reactor sales in developing areas have been arranged on a more-or-less turnkey ("package deal") basis with foreign contractors, it is possible that under present circumstances, developing countries could obtain bargains from leading nuclear vendors.

All in all, local cost factors, problems in assuring a fuel supply, plus the health and environmental effects and local technical infrastructure problems already mentioned, should all be given full weight in assessing the desirability of nuclear power in developing-country planning.

From Biomass Combustion

Compared with many of the other solar and renewable technologies, the burning of wood to heat boilers is a well-established technology and is currently used in many developing areas. Some advances have been made in industrial countries in the use of wood and wood wastes, in particular in methods of harvesting and compacting the fuel. Rough estimates would indicate that wood-burning generation of electricity is more or less competitive with coal and nuclear power in the United States—for wood prices that appear feasible in particular geographical areas.[22] The relative advantages of this technology in many developing areas—where wood in some locations is not fully harvested and the total biomass resource is large—could be even greater.

[22] Schurr and coauthors, *Energy in America's Future*, pp. 316–318.

The question requires more study and investigation through pilot projects. In particular, the more efficient use of wood and wastes in many developing areas might have to deal with countervailing social pressures from present uses as a traditional fuel or fertilizer. Some planners in Latin America, for example, recently have stressed schemes for direct combustion of wood in small *leñoeléctricas* that utilize only presently unharvested wood in forest locations far from a conventional electric grid. But the advantages of an established technology, such as the combustion of wood, plus the underutilization of many of the world's forests, points to a promising role also for large-scale plants supplying wood-generated electricity to central grids.

From Geothermal Energy

It is well established that power from dry steam sources can be competitive with fossil or nuclear electricity.[23] Table 8-1 shows some capital costs estimates, and unit costs of 15 to 30 cents per kilowatt hour for the still somewhat experimental use of liquid-dominated sources have been calculated.[24] As is the case with many new sources of energy, however, the geothermal resource base has not been fully investigated. But as emphasized in chapter 7, experience in a few countries would suggest that dry steam resources are limited in extent. The use of liquid-dominated geothermal fluids and other geothermal possibilities are for the most part still in the testing stage; however, progress in this field should be watched closely. Should current investigations in the United States and elsewhere prove fruitful, geothermal options for developing areas should receive close attention.

From Other Sources

Energy for central-station electricity can come from other renewable sources: direct solar, ocean thermal and wave power, tidal energy, and wind energy. With the possible exception of wind, all of these technologies are either still in an experimental stage or exceedingly expensive using present techniques. In particular, solar photovoltaic systems could contribute to central-station generation, but they promise to remain rather expensive during the next decade at over 7 cents per kilowatt hour in U.S. applications, even taking an optimistic view

[23] Ibid., pp. 319–321.
[24] Ibid., p. 321.

about anticipated advances in collector cell technology.[25] It is well known that wind energy generators work reliably on at least modest scales. More economic large-scale generators can be expected to reach costs approaching 2 cents per kilowatt hour for average wind speeds in the range of 25 to 30 kilometer per hour; however, they will probably require more testing before adoption by most developing countries.[26] Furthermore wind, like many of the solar-based sources, suffers from serious interruptibility problems, with consequent need for expensive storage for most applications. Some of these renewable technologies may be more promising in a decentralized context rather than as sources of central-station electricity. But the role of scale is complex. (See "Village Technologies".)

Transmission and Electricity Scale Costs

The prospect of changes in the fuel mix for electricity brings into focus the important role of transmission costs for electricity—and therefore, what the optimum scale of new generating plants should be. While it is difficult to generalize about costs for all developing areas, estimates of roughly 0.5 cents per kilowatt hour have been made for average hauls of some 340 kilometers.[27] One-half cent is not a negligible cost. But typical marginal savings in transmission costs that could be gained by making small decreases in the average transmission distance through decentralization of generating plants would be expected to be much lower. And, although table 8-1 does not show total costs per kilowatt hour, it suggests that economies of scale could in many cases override such transmission cost savings of, say, 10 percent or even less. Transmission losses also have to be taken into account—they raise costs by another 10 percent on a national average in the United States and 20 percent or more in some countries. But again, reduction of scale would only save a fraction of this cost also. In sum, the reason for the historical trend to larger plants is then evident: capital costs per

---

[25] Ibid., pp. 308–309. But optimism of U.S. Department of Energy solar planners appears to grow with time. See costs of 4 cents per kilowatt hour hoped for in 1986 (U.S. Department of Energy, *National Photovoltaic Program, Multiyear Program Plan* [June 6, 1979] table 2a, p. xii).

[26] Schurr and coauthors, *Energy in America's Future*, pp. 311–313.

[27] From Oak Ridge Associated Universities (ORAU), *Future Strategies for Energy Development,* Proceedings of a conference at Oak Ridge, Tennessee, October 20–21, 1976 (Oak Ridge, Tenn., ORAU) pp. 115n, 116. One can deduce approximate separate transmission and distribution costs from residential and industrial totals.

kilowatt of capacity can be decreased 40 percent by tripling the plant size.

These conclusions are strictly valid only for load factors typical of the United States. For smaller usage patterns, smaller scale plants closer to population centers—like small-scale hydro—could be desirable. Furthermore, such conclusions also only hold for readily transportable fossil fuels (or for no fuel at all—as in the case of hydro plants). If more bulky biomass sources are used, scale questions may have to be rethought.

Finally, if potential new sources are very far from load centers, thought may have to be given to entirely different alternative energy sources—or if the economics allow—to transmitting electricity on direct current transmission lines, which are cheap per energy unit if loads are sufficiently large.

Alcohol Fuels

The need for liquid fuels, especially fuels suitable for internal combustion engines, is of the highest priority for developing areas as well as for the industrial countries. Nearly 67 percent of total commercial energy consumption in developing countries consisted of petroleum and refined products in 1976, as compared with only 51 percent in the industrial countries.[28] While about half of this consumption might be replaceable with other nonliquid sources for generating electricity or providing mechanical energy, there is at present no viable alternative to liquid fuels in most of the transport sector. Such fuels can be produced from a number of sources, including coal; but given the prospect of much higher prices for oil and gas, and the restricted availability of coal in many countries, the role of biomass in supplying liquid fuels is a major focus of interest.

Special attention in this connection has been devoted to alcohols, particularly ethanol (or grain alcohol) and methanol (or wood alcohol). Production technology for the alcohols is either established or relatively close to practical use. Ethanol is produced from sugars, starches, or cellulosic materials through fermentation; methanol is produced from wood using thermochemical conversion (heating) techniques. Pilot methanol facilities and commercial large-scale ethanol operations exist,

[28] United Nations, *World Energy Supplies 1972–1976* Series J, No. 21 (New York, 1978).

with the Brazilian use of a 20 percent ethanol-gasoline mixture in automobiles being the best known and most ambitious program to date. Since in theory fuel alcohols can be produced from any living matter, the resource is potentially large, even though as pointed out briefly in chapter 7, the problem of resource competition—fuel versus food or forests—is quite complex.

Technologies

Two basic types of technology exist for producing ethanol. If a feedstock that is high in sugars is used, such as sugarcane, the sugars can be fermented using a yeast enzyme to produce carbon dioxide and ethanol and subsequently distilled to concentrate the alcohol. If starchy or cellulosic materials, such as food grains and wood, respectively, are used for ethanol production, they must first be turned into sugar by treatment with heat and enzymes, and subsequently into alcohol. The process for grains involves the familiar technologies used in the production of beverage alcohol from corn, wheat, and other grains. For wood, no commercial process exists. Still, the possibility of using wood as a feedstock is of great interest because of its lower market or opportunity cost value relative to that of sugars and starches as foodstuffs. Plants using an acid hydrolysis process to break up long cellulosic carbohydrate molecules into shorter sugar carbohydrates and thence into alcohol have been built in the United States, the Soviet Union, and Switzerland and are now being designed in Brazil.[29] Enzymes might also be used to extract sugar from wood; this method is still at a pilot stage.[30]

Methanol is one of the end products of the thermochemical conversion of the constituents of biomass into new compounds that are convenient for many energy end-uses. These processes involve breaking the chemical bonds of the organic molecules by applying heat or high pressure. A synthesis gas derived in this manner from heating biomass can be turned into liquid methanol by means of a long-

[29] There is also a large body of related experience with the chemistry of the process, since the decomposition of wood by acids or alkalis is routinely carried out as part of the operation of making wood pulp, and sugars and alcohols are a common by-product (A. J. Panshin, E. S. Harrar, J. S. Bethel, and W. J. Baker, *Forest Products: Their Sources, Production, and Utilization* [New York, McGraw Hill, 1962] p. 428ff).

[30] Steven F. Miller and Jackson Yu, "Production of Ethanol from Lignocellulose—Current Status. Paper presented at the Third International Symposium on Alcohol Fuels Technology, Asilomar, Calif., May 28–31, 1979.

established, completely commercial technology. This process involves first "shifting" the hydrogen-to-carbon ratio to the higher value characteristic of the methanol molecule and then combining them by using a catalyst to "reassemble" the gas molecules,[31] creating a crude methanol that can be further refined or used directly as fuel.[32]

Uses and Constraints

The principal use of alcohol fuels of interest for energy policy makers is as a replacement for gasoline and higher petroleum distillates, in particular as a motor fuel. But other possible uses include replacement of residual oil under boilers and as hydrocarbon fuels in gas turbines.[33] Alcohols can also be used to replace kerosine in lighting and cooking, especially in rural areas.[34] Use in internal combustion engines can be either in a relatively small percentage (10 to 20 percent), in a gasoline-alcohol blend in standard engines, or as pure alcohol in a modified engine.

The virtues and drawbacks of ethanol and methanol as a motor fuel appear similar. Both increase the fuel octane rating: methanol is already frequently used as a fuel in racing cars. Although one study has maintained that the mileage increases for gasohol were not statistically significant,[35] evidence has been presented that about 7 percent less fuel in energy terms is used by vehicles in tests with gasohol containing ethanol.[36] Other calculations have claimed that pure ethanol—in contrast to gasohol—in properly designed engines is volume for volume as efficient as gasoline, or 45 percent more efficient in gross energy

[31] In fact, commercial methanol is made by first "reforming" natural gas into synthetic gas (shifting the carbon-hydrogen ratio back the other way). A possible alternate to the production of pyrolytic gas, cleaned biogas from anaerobic fermentation, could also be used as a hydrocarbon source to replace natural gas feed; this possibility has been little examined.

[32] For industrial use, this 75-to-90 percent crude methanol product is then further processed, but such processing may not be necessary for its use as fuel.

[33] David LeRoy Hagen, *Methanol: Its Synthesis, Use as a Fuel, Economies, and Hazards* (Washington, D.C., Energy Research and Development Administration, 1976) p. II-46.

[34] Pure alcohols cannot be used in diesel engines because of vaporization-point problems, but blends could be.

[35] Jerry R. Allsup and Dennis B. Eccleston, "Ethanol-Gasoline Blends as Automotive Fuels." Paper presented at the Third International Symposium on Alcohol Fuels Technology, Asilomar, Calif., May 28–31, 1979.

[36] William A. Scheller, "The Use of Ethanol-Gasoline Mixtures for Automotive Fuel" (Orlando, Fla., Institute for Gas Technology, 1977) p. 194.

terms.[37] Undoubtedly, further test experience on a variety of engine types and under different driving conditions is needed to resolve these questions.

There are certain problems in operating automobiles on a blend of alcohol and gasoline. Alcohol used in such blends must undergo, at some extra cost, a final processing step to remove all water, or a phase separation—like oil and water—may take place that can impede engine operation. Because of the effectively higher octane rating of alcohol, the leanness of the mixture could cause stalling and hesitation; careful matching of the octanes in the blend should limit this problem. Vapor lock is a potentially more serious driveability problem, since the alcohol in a blend could vaporize at a different temperature than the gasoline hydrocarbons and impede fuel feed.[38]

Using pure methanol or ethanol in engines would eliminate some of these problems, but specially designed engines would have to be built. Carburetors would have to be modified for different air-to-fuel mixtures, and boiling point problems might necessitate the use of fuel additives or a provision for heating the fuel intakes. Pure methanol also has a tendency to attack certain plastics and metals—in particular the tin-lead mixture often used to coat gasoline tanks—but replacement materials are available.[39] Toxicity effects of methanol also require study (see chapter 9).

Costs

The costs of ethanol are somewhat better established than those of methanol, since it has been produced widely on a commercial scale. However, proposed methanol schemes would combine two relatively well-known engineering elements: the production of synthesis gas from organic materials and the methanol production ("methanolization") process itself. Furthermore, ethanol for fuel use may involve new

[37] Thomas A. Sladek, "Ethanol Motor Fuel and Gasohol," *Mineral Industries Bulletin* vol. 21, no 5 (Colorado School of Mines Research Institute, May 1978) p. 5.

[38] American Petroleum Institute, *Alcohols: A Technical Assessment of Their Application as Fuels,* Pub. No. 4261 (Washington, D.C., July 1979) p. 8; also, David LeRoy Hagen, *Methanol,* p. II-35.

[39] G. R. Cassels, W. G. Dyer, and R. T. Roles, "BP New Zealand Experience with Methanol/Gasoline Blends"; E. Earl Graham, Barry T. Judd, and Vivian Alexander, "New Zealand's Methanol-Gasoline Transport Fuel Program"; and A. Svahn, "Methanol/Gasoline Mixtures in Four Stroke Auto Engine." Papers presented at the Third International Symposium on Alcohol Fuels Technology, Asilomar, Calif., May 28–31, 1979.

feedstocks or improved technologies for mass production for non-chemical purposes.

Ethanol. In fact there have been several studies of ethanol costs per unit of energy. These are summarized in appendix table 8-A-1. The values in the table, given in gigajoule units, range from the equivalent of about $500 to $800 per metric ton oil equivalent (toe) for sugarcane and anywhere from $400 to almost $1,000 for wood feedstocks, with similar results for corn and manioc (cassava).[40] In terms of 1975 dollars, even the lower end of this cost range is higher than prices of gasoline from imported oil in the United States in early 1980 of about $350 per metric ton oil equivalent.

Ethanol costs appear to be relatively sensitive to feedstock costs, and less to capital costs.[41] This may, however, be less true for feedstocks that have significant by-product credits, such as wood and corn. Appendix table 8-A-1 gives some costs for various types of feedstocks in dollars per gigajoule, illustrating their relative importance in contributing to total costs.

It is also evident from this table that by-products may be important in the economics of ethanol production. For sugarcane, the bagasse (or waste from sugarcane milling) is often used in developing areas as a source of heat for process steam and distillation; these credits are already implicitly included in operating and maintenance costs and are not shown in the table. Cane by-products can also be used as structural materials and animal feeds. On the other hand, disposal of by-products can also represent an additional social cost because of potentially severe effects of organic wastes from the bagasse on local aquatic ecosystems.

For production of ethanol from grain, the spent grain residue resulting from the waste mash is presently a commercial product that could have a fairly high value in future markets, potentially reducing the feedstock costs by about one-half, as shown in appendix table 8-A-1.[42] By-products from the conversion of wood to sugars consist of

[40] The values shown here are roughly consistent with the results of a recent review of a number of studies that showed a range of estimates of $500 to $800 per toe for ethanol from sugarcane and sugar juice and $800 to $1,100 per toe from wood (derived from MITRE Corporation, Metrek Division, *Comparative Economic Assessment of Ethanol from Biomass,* prepared for U.S. Department of Energy, September 1978 [available from National Technical Information Service HCP/ET-2854] p. 104).

[41] MITRE Corporation, *Comparative Economic Assessment,* pp. 111–113.

[42] Sladek, "Ethanol Motor Fuel," pp. 10–11.

useful chemicals such as methanol and furfural—used in plastics manufacturing—and of lignin residues, which can be used as a fuel to provide steam for the process itself or as a plastics feedstock.[43]

Possibilities of reducing costs of the phases after the sugar is formed exist, but from an internal energy standpoint the final process of turning sugar into alcohol is quite efficient: efficiencies of 80 to 90 percent are easily obtainable.[44] The key problem with ethanol costs is that sugar *of any kind* tends to be an expensive commodity—even, by inference, the raw, unrefined sugars produced from starches and wood.

Methanol.   Methanol cost estimates shown in table 8-A-2 appear to vary a good deal. But if one considers only methanol produced from biomass feedstocks that are available in a given cost range, the table shows that there may be more uniformity than is immediately apparent. The table shows that for wood or other biomass materials available at $40 or less per toe (about $1 per gigajoule), costs range from about $200 to $300 per toe (about $4 to $6.50 per gigajoule) of energy in the form of methanol.

Prices in the existing commercial market for methanol produced as a petroleum by-product have fluctuated widely in the past decade, perhaps partially explaining why cost estimates for fairly large plants to convert wood to methanol vary greatly as well. On general grounds, a total of $170 per toe for methanol from a $40 per toe biomass input is a reasonable lower bound on costs for technology that can be expected for the foreseeable future.[45]

The future costs of biomass inputs are of course of critical importance. Forty dollars per toe for wood or wood waste is roughly consistent with data from several locations,[46] but the biomass resource cost question requires more study.

[43] Panshin and coauthors, *Forest Products*, pp. 490–493.

[44] Raphael Katzen Associates, *Chemicals from Wood Wastes* (Cincinnati, Ohio, December 1975) p. 74 (available from the National Technical Information Service, PB. 262-489).

[45] A lower bound on unit costs can be derived by examining the cost of the input feed and the efficiency of the process. Assuming a slightly higher efficiency than present practice of about 50 percent, the base cost of methanol from $45 per toe of wood will be at least $85 per toe of alcohol in terms of costs of biomass and feedstocks only, plus whatever capital and operating costs apply. Based on capital and operating costs for converting natural gas feed into methanol, a minimum of perhaps $85 per toe for total other non-fuel costs would be a likely estimate.

[46] See chapter 6 in this book; Elizabeth Cecelski, Joy Dunkerley, and William Ramsay, *Household Energy and the Poor in the Third World* (Washington, D.C., Resources for the Future, 1979) chapter 3; and Hagen, "Methanol: Its Synthesis," section 3.

Factor Cost Feedback

Much concern has been expressed that quoted costs involve potentially disastrous feedback effects. For example, future use of expensive fluid fuels in agriculture or sugar or alcohol processing could conceivably vitiate present-day calculations for alcohol production at $500 per toe or so. That is, if the energy needed to power phases of production also had to be provided by high-cost alcohol liquids, instead of by gasoline or natural gas as at present, the *net* production of alcohol fuels could be very low—and alcohol therefore very costly. This concern is more cogent for industrialized countries than for developing areas: in the United States natural gas is used in distilling, but in Brazil, bagasse is utilized. Furthermore, even when liquid fuels are now used for tractors or to manufacture fertilizers, price elasticity and substitution in these productive processes as fuel costs rise in the future must be expected—leading to a reduction both in the quality and the quantity of the energy needed to produce alcohol.[47]

Thus, this point about feedbacks, while well taken, must be more closely studied for its implications for alcohol fuel production in industrial countries. Given the much smaller direct and indirect inputs of fossil fuels and the agricultural and alcohol production technologies actually used in many developing countries, the "cost feedback," sometimes called the "energy balance" effect there can be expected to be far less important.

Land Use Implications

Evaluation of land use requirements and costs is critical not just in the production of liquid fuels from biomass, but for all types of biomass fuels. In fact, the value of land used to produce biomass for fuel alcohols will increasingly be compared to its value in producing biomass for other fuel uses, such as direct combustion to produce steam or electricity, as well as in producing alternative nonenergy products, such as pulp, timber, or agricultural crops.

In developing countries, the most serious potential competitor with biomass for land is likely to be food production; like other cash crops, high-value energy crops—particularly if supported by guaranteed prices

---

[47] Elizabeth Cecelski and William Ramsay, "Prospects for Alcohol Fuels in Developing Areas." Paper presented at the UNITAR Conference on Long-Term Energy Resources, Montreal, November–December, 1979 (Washington, D.C., Resources for the Future).

or purchases—could displace food production and hence drive food prices up. Alcohol production from wood feedstocks, or from other crops which can grow on less desirable land (or water), are less likely to compete with food production.[48]

The amount of land needed to produce alcohol fuels can vary greatly, depending upon yields, sugar content of the feedstock, conversion, and end-use efficiencies. Appendix tables 8-B-1 and 8-B-2 have illustrative estimates of how much land some of the larger producers of potential feedstocks might have to devote to energy crops or forests in order to meet current liquid fuel requirements, assuming current national agricultural yields and several alternative yields for forest biomass. These estimates are highly speculative: for example, future yields for crops such as sweet sorghum, which is not presently widely grown commercially, are tremendously difficult to predict. On the other hand, huge possibilities for improvement in yields for conventional crops often exist. To date, most genetic improvement has been directed at maximizing specific forest or food products, not total biomass volume or energy content. But high yields are now often achieved through energy-intensive fertilizer, machinery, and irrigation inputs. The extent to which labor could be substituted for these inputs is another important topic for investigation in the labor-surplus developing countries.

Besides improving yields, another way to reduce competition of energy biomass with food production is to grow biomass on marginal lands that are less suitable for food production. It is clear from table 8-B-1 that sugarcane is the highest producer of alcohol per hectare, given current yields; sweet sorghum, cassava, and wood, however, may be cultivable on lower quality land and thus be more desirable from a land use point of view. Nonetheless, most field crops naturally produce far better yields on prime lands with less energy and other inputs than on marginal lands. The critical question, then, for both agricultural and forest energy crops still tends to center on the possible yields from lands that will turn out to be available for energy crops in each particular geographical region.

It should be noted that appendix tables 8-B-1 and 8-B-2 do *not* consider costs of cultivation and conversion, or whether the quality or quantity of land, water, or climate required would actually be available

---

[48] Crop and forest residues and aquatic biomass are other potential feedstocks that have not been considered here, but that may be of future importance (see chapter 6).

to produce the quantities indicated.[49] But it is interesting to see that some countries with low liquid fuel requirements relative to their available land areas (such as India, Indonesia, and Ethiopia) appear to be capable—in a very theoretical sense—of fulfilling their liquid energy consumption from biomass by utilizing a relatively small part of their total available arable or forest land, particularly if indicated wood yields turn out to be achievable on a large commercial scale. For other countries with large liquid fuel consumption relative to their available land (such as the United States, Egypt, and Cuba), alcohol fuels would at best only be able to meet a portion of their liquid fuel requirements. Nonetheless, a more detailed examination of the suitability and availability of actual and potential crop lands, as has been partially done for some crops in the United States and Canada, would be necessary in order to draw more concrete conclusions.[50]

"Village Technologies"—Energy Systems Appropriate to Rural Use

The need for larger quantities and higher qualities of energy resources for rural economic development is critical. Supplies of traditional fuels—wood and wastes—are already under strain in many areas (see chapter 2). The "village technology" approach is in some sense a strategy for exploring nontraditional and more efficient ways of utilizing these traditional resources. Solar cookers, for example, may use a nontraditional device, the reflecting paraboloid. On the other hand, the solar energy resource itself is patently traditional.

Village technologies are one example of a class of so-called appropriate technologies. Appropriate technologies are, in general, technologies that are fitted to the available resources of the energy user—in the sense that "resources" means all human and material factors.

---

[49] It would also be useful to estimate what the land use implication would be of replacing only gasoline, rather than all liquid fuels, with alcohol fuels. Most interest has been shown in this partial substitution, and there is some evidence that conversion efficiencies of alcohol when used in internal combustion engines may be higher than gasoline. Therefore, the energy replacement potential of alcohols may be higher than indicated in the table, where standard "lower heating values" are used.

[50] E. S. Lipinsky, S. Kresovich, and coauthors, "Systems Study of Fuel from Sugarcane, Sweet Sorghum, and Sugar Beets," third quarterly report to the Energy Research and Development Administration (Columbus, Ohio, Battelle Columbus Laboratories, April 14, 1976); and Inter Group Consulting Economists, Ltd., *Liquid Fuels from Renewable Resources: Feasibility Study* (Winnipeg, Manitoba, March 1978).

Current interest in such a "resource fit" is high, perhaps because in part smaller scale and often technically simple technologies[51] may have been unfairly neglected in the past and may represent more truly economic alternatives.

The other aspect of current interest in appropriate technologies has to do with the more convoluted relationships of technologies to existing life-styles, to the environment, and to the development and cultivation of participatory democracy and other hard-to-quantify concepts. For example, promoters of village energy technologies commonly try to make certain that the technologies can be produced at the local level or at least that their use emphasizes local control and participation. This last aspect, involving at times such complex concepts as avoidance of alienation and stimulation of participatory democracy, could quite conceivably produce substantial benefits to the community.[52] These noneconomic benefits will be briefly discussed below, but first we will consider the possibility that village technologies, such as wind generators and mini-hydro facilities, can satisfy end-uses on a cost basis that is competitive with conventional sources such as diesel pumpsets and rural transmission lines. Furthermore, these technologies may have some interesting macro-economic effects that should be considered.

Economics

The purely financial viability of village technologies can be directly compared with alternatives, such as rural electrification, in terms of dollars and cents. In spite of the fact that many of these technologies are new and costs cannot yet be estimated in a very reliable manner, it is possible to gain a general idea of relative costs.

Direct solar technologies. In many developing areas, perhaps the most promising application of direct solar technology—the use of the sun's rays as heat or in photovoltaic cells—is in the use of flat plate collectors for crop drying. Many crops are at present dried in the sun, and solar flat plate collectors should be relatively inexpensive if they are to supply the low-temperature heat characteristic of drying proc-

---

[51] Note that some technologies often considered "appropriate" are only simple to use, not to make, for example, photovoltaic cells.

[52] William Ramsay and Elizabeth Cecelski, "Energy, Scale, and Society" (Washington, D.C., Resources for the Future, 1980).

esses. Indeed, one assessment has estimated that in Africa, active solar systems, especially if aided by the incorporation of passive features, could substitute in some cases for drying by using oil and wood. However, the same review points out that crop drying typically takes place by natural means, and there is very often no incentive to go to more expensive systems, whether fuel-fired or solar.[53]

Using solar thermal engines, which focus the sun's rays to heat water or some other fluid to make a steam vapor for irrigation or electricity generation, does not seem promising, both because of the initial expense and the necessity for maintenance. The case of photovoltaic cells—where sunlight is turned directly into electricity—is somewhat more complicated. While costs are presently high, optimism about future lower cell costs is great.[54] Even at recent costs of about $10,000 per peak kilowatt (that is, the cost of cells to produce one kilowatt when conditions are optimal), electricity could probably be produced at approximately $1 per kilowatt hour.[55] This, while expensive, is usually cheaper than batteries for isolated operation of such appliances as television sets or refrigerators. Storage of energy is of course a basic problem that must be faced: irrigation uses where water pumping can be delayed until bright daylight conditions occur are therefore particularly attractive.

Increasing experience in the use of photovoltaics in isolated applications in developing areas should help in proving out the reliability of the system under demanding conditions. In contrast to a solar thermal system, proponents hope a photovoltaic pump will have fewer moving parts and function with relatively little maintenance. Given high oil prices or long distances to electric generating stations, such pumps—if successful—would then be extremely useful.[56]

Much less promising is the solar cooker. Admittedly, costs for these cookers may be relatively "low" (typically $7 to $35).[57] A $10 cooker

[53] Julia C. Allen, "Economic Aspects of Solar Grain Drying of Food and Export Crops in Africa," Unpublished research paper, May 8, 1979, p. 23.

[54] See U.S. Department of Energy, "National Photovoltaic."

[55] U.S. National Academy of Sciences–Tanzania National Scientific Research Council, *Workshop on Solar Energy for the Villages of Tanzania*, Paper presented at a conference held in Dar es Salaam, Tanzania, August 11–19, 1977 (TNSRC, 1978) p. 37.

[56] In a project which has had extensive support from abroad, William Bifano reports good results with local volunteers as operators, although reportedly the program has experienced some problems. (William Bifano, "Description of a Photovoltaic Power System in the Remote African Village of Tanzaye, Upper Volta." Paper presented at the UNITAR Conference on Long-Term Energy Resources, Montreal, November–December, 1979).

[57] Cecelski and coauthors, *Household Energy*, pp. 12, 54.

could replace a good part of the energy from traditional fuels used by a family at costs of only about $4 per toe.[58] Still, this option suffers from being a replacement for traditional fuels where these often present an undetermined but perhaps extremely low opportunity cost to the rural villager. Furthermore, problems in acquiring such capital items—even on the modest scale mentioned here—and cultural obstacles in using such new appliances can be prohibitive. These problems will be discussed in chapter 9; but it can be noted here that a cooker that operates only outdoors, during fair weather, in the daytime, and on only one utensil at a time cannot be expected to supply the exact socioeconomic equivalent of a campfire or primitive wood stove.

Wind energy.   Small windmills for pumping water or wind generators for electricity are an established technology. In addition, research has been carried out on adapting local materials to modern forms of this technology. Small generating units have existed on a commercial basis for some time and could possibly supply electricity in some locations. Even so, such electricity is usually rather expensive unless the windmill is tied into a large electric grid—which generally takes it out of the realm of village technology (see above).[59]

Of even greater concern is that in developing countries, and especially in tropical areas, average wind velocities are often low—an aspect that is more serious than is sometimes apparent, because wind power decreases as the cube of the wind speed. Without reasonably priced storage facilities, wind-generated electricity may also be ill adapted to many end-uses requiring a constant electric current. Nonetheless, water-pumping technologies, where some interruption in supply is less important, will likely see an increase in the use of wind power as oil prices rise.

Small-scale hydroelectric power.   Depending on the site, small-scale hydroelectric power can provide electricity at prices that will be competitive in many developing-area contexts, and although economies of scale hold for hydroelectric plants as for most electricity technologies, special equipment such as bulb turbines designed for low heads[60] can improve the economies of small-scale operations. In U.S. planning, the use of existing dams built for nonpower purposes is also an

[58] Ibid., p. 12.

[59] Schurr and coauthors, *Energy in America's Future*, compare pp. 311–313 and pp. 332, 335.

[60] "Head" is the height of water in the reservoir.

important way to save money, but this factor is often not important in other countries. If dam costs are included, it seems reasonable to believe that electricity can be delivered at costs of 5 to 10 cents per kilowatt hour.[61] If flows are not excessively seasonal, this renewable source could be competitive with other electricity sources in many rural areas. It is not clear, however, that the fairly advanced degree of technical sophistication and maintenance required for this technology is fully consistent with some of the participatory benefits associated with village technologies mentioned earlier. But as a logical extension of normal hydroelectric planning, small hydroelectric plants will undoubtedly become an important adjunct to energy technologies in developing areas.

Biogas. Biogas produced by fermenting biomass materials in an enclosed space appears to be a very promising technology for some countries. The economic outlook is complicated by differing claims and by the patterns of subsidies used in some countries to support this technology; nevertheless, prices of $30 to $130 per metric ton oil equivalent have been quoted.[62] However, if either animal dung or vegetable wastes are used as the input, the costs of alternative uses of the wastes as fertilizer or fuel must be taken into account. In any event, the fertilizer value is mainly in the nitrogen content, and a very large proportion of the nitrogen in the original wastes is preserved in the process of fermentation to produce the biogas. Indeed, the value of the residue or slurry from biogas generators is often high in local village markets. The energy itself in the waste also often has a low opportunity cost: its low-grade fuel value in inefficient burning in open fires or crude stoves. But complex social constraints may have to be taken into account (see chapter 9).

As with all relatively expensive capital items, credit and equity problems must be taken into account. We also discuss this in chapter 9. Furthermore, this technology is not suited to every country, either because of the lack of a suitable waste resource base or because climate conditions are too cold for effective anaerobic fermentation. The general applicability of biogas may depend on future success with vegetable wastes; although all biomass wastes can be anaerobically fermented in theory, in practice there is a considerable advantage in the "preprocessed" qualities and homogeneity of animal wastes.

[61] Cecelski and coauthors, *Household Energy*, pp. 46–48.
[62] Ibid., pp. 39–40.

Although most biogas units in India, where the technology has been especially favored, have been built on an individual basis, many industrial-size and collective plants have been built in other countries.[63] It is to be expected that economies of scale would also hold for this technology, and that larger installations might be an especially useful option in many developing areas. In some cases such a larger scale technology might still fit into the village technology framework. In other cases, a commercial or industrial format might be a more appropriate option for the local economy.

"New" wood and charcoal technologies.   Much of the traditional use of wood as a cooking fuel in developing areas is carried out in campfires or stoves where the efficiency of heat use is very low. As already discussed in chapter 5, that means that an obvious way to increase the amount of energy available is to introduce improved cooking stoves. Merely by improving technical features, draft control, for example, efficiencies often can be tripled.[64] Since the technological, and especially the social changes involved, should be small compared, for example, with solar cookers, government aid may be particularly appropriate in this area to overcome credit barriers to the use of improved fuelwood or charcoal stoves.

In considering such changes, the possibility for replacing fuelwood use with charcoal should be considered. Charcoal is cheaper to transport and easier and more convenient to burn. This replacement has tended to occur historically in developing countries, when transport distances became long or when incomes rose, and especially in urban areas. If the charcoal is made in modern kilns, the overall efficiency (charcoal making plus charcoal burning) of using charcoal can be only slightly less (or even greater) than that of using fuelwood, so energy efficiency goals can also be pursued by this strategy. Again, the provision of high-cost kilns, for example in excess of $1,000, can involve a credit problem.[65] In many countries, charcoal making has

[63] Andrew Barnett, Leo Pyle, and S. K. Subramanian, *Biogas Technology in the Third World: A Multidisciplinary Review* (Ottawa, International Research Development Center, 1978).

[64] Philip F. Palmedo, Robert Nathans, Edward Beardsworth, and Samuel Have, Jr., *Energy Needs, Uses, and Resources in Developing Countries*, report to the U.S. Agency for International Development (Upton, N.Y., Brookhaven National Laboratory, March 1978).

[65] David French, "Economic and Social Analysis of Renewable Energy Projects: The State of the Art." Paper presented at the U.S. Agency for International Development Africa Bureau Firewood Workshop, Washington, D.C., June 12–14, 1978, p. 12.

traditionally been left to small-scale local entrepreneurship, and kiln improvement could therefore require major institutional changes.

These improvements in wood and charcoal technologies of course only make sense if the forestry resources involved are used wisely. The need for preventing deforestation is well known. Nonetheless, the complementary requirement to manage present forestry resources in an intelligent manner so as to yield a sustainable supply of biomass fuels for both the traditional and modern areas should be equally emphasized. Careful planning in growing, harvesting, and transportation can pay big dividends in helping to increase the effective energy supplies of many developing areas.

Macroeconomic Effects

Even if some village technologies should not turn out to be economic in the sense of dollars spent, it may be that other macroeconomic effects can compensate. Particular effects that are of policy interest are employment, especially rural employment, and related questions of regional and income-class equity. Land use planning goals, such as the discouragement of rural migration to cities, could also be called on to counterbalance any possible microeconomic disadvantages.

Such macroeconomic aspects of village energy technologies are the same kind that have been key factors in formulating certain other development strategies in the past, notably integrated rural development policies. In many rural-oriented development schemes, the cost of increasing employment, for example, may be high but yet willingly incurred by some planners in order to achieve specialized employment goals. Village technologies, or other energy technologies that are designed to be particularly appropriate to a given region, may contribute to that objective. All the same, there is no immutable connection between the "appropriateness" of village technologies and most macroeconomic factors: larger scale energy technologies such as rural electrification may sometimes be the most effective way to produce desired regional economic effects.

Sociopolitical Factors and Energy Scale

Some social critics would consider a discussion of the economic merits of village technologies an inadequate response to the societal questions involved. To be sure, there are always factors that are omitted in a purely dollars-and-cents analysis; the environment is a well-known example that has concerned analysts in recent years (see chapter 9.) In this connection, a recent criticism of existing energy technologies[66] has included not only environmental factors, questions of reliability, and hidden inefficiencies in large enterprises, but also those questions of wider scope which may be called "sociopolitical."

Sociopolitical factors enter the picture because village and other small-scale technologies might help to provide such complex benefits as a stimulus for cooperative community life, fuller utilization of individual capabilities, and, in general, a lessening of the symptoms of social alienation that are thought to be typical of an industrial society. Explicit political results, such as the development of institutions encouraging participatory democracy at local levels, could be considered a key benefit of appropriateness of energy supply schemes. Indeed, much of the literature on this subject pursues themes on social cooperation and politics by consensus that are similar to those developed in earlier anarchist literature.[67]

The ramifications of this controversy are rather complex.[68] But the most important sociopolitical criteria for deciding between the adoption of village and nonvillage technologies for energy purposes will revolve around the values that a local population may wish to attach to the ability to participate in common decisions and tasks, do-it-yourself projects, and community activities in general. Cooperative decision making and individual self-reliance in handling work tasks and social problems are concepts that are highly prized in many societies. Be that as it may, these goals usually cannot be pursued without some kind of attendant social cost. In assessing the value of these sociopolitical goals in the context of energy applications, one must also include in the balance the drain on money, time, and other resources that must be spent in operating and taking care of new energy technologies, for example, community biogas plants. It is clear that there must be disadvantages as well as advantages in widening responsibilities and decentralizing the control over technologies. The questions of who is to run the community biogas plant, of how dung will be collected, of whether the gas should be metered or not, are all examples of problems

[66] See especially Amory B. Lovins, *Soft Energy Paths: Toward a Durable Peace* (Cambridge, Mass., Ballinger Publishing Co., 1977).

[67] P. A. Kropotkin, *Selected Writings on Anarchism and Revolution* (Cambridge, Mass., MIT Press, 1973). Ivan Illich, *Tools for Conviviality* (New York, Harper and Row, 1973).

[68] See Ramsay and Cecelski, "Energy, Scale, and Society."

that must be faced. Social scientists recognize, for example, that not everyone appreciates increased responsibility for social tasks or the necessity for sometimes prolonged social interaction involved in participatory decisions.

If a strategy of seeking these sociopolitical benefits from village technologies might be characterized in somewhat exaggerated form as "energy systems for the sake of social cooperation" instead of "social cooperation for the sake of energy systems," then the question becomes: What price social cooperation, community participation, and the building of individual self-reliance? There is no hard and fast way to answer this question on the basis of the meager store of social data available. But energy planners in developing areas must necessarily make value decisions about the desirability of these sociopolitical effects if they are to fully assess the role of some of the relatively expensive appropriate-scale village energy technologies. It should also be emphasized that these kinds of judgments are generally best left to national and local authorities; outsiders from industrial countries usually have no special qualifications to contribute to them.

Conclusions

This chapter has reviewed two critical energy supply areas—electricity and liquid fuels from alcohol—and has examined the village (appropriate rural) energy technology option that is of great current interest in rural development planning.

After considering all factors, the outlook for substitute fuels to replace imported oil in electricity generation in developing areas is troubled but has some bright spots. The present state of the nuclear industry confronts planners with serious uncertainties. On the other hand, coal is promising if it can be found locally or imported at reasonable prices.

One solution for electricity generation is to develop renewable sources, particularly hydroelectric power. In many countries, although the best hydro sites usually have already been used, formerly marginal sites have received greater attention since 1973 and should continue to do so.

Apart from this alternative, many countries must give serious consideration to new options for producing electricity, perhaps geo-thermal, but in particular, the widespread resources available for biomass conversion. Large generating plants of this type have not yet been built, but the known technical and relatively reasonable costs of existing small plants must encourage planners to carry out a thorough investigation of this possibility for electricity generation.

Use of biomass, which is expensive to transport, and of remote hydroelectric resources far from load centers will involve the transport of either fuel or energy over long distances. Therefore, energy planning for these resources may have to reexamine closely questions of plant scale and transmission line alternatives.

Ethanol and methanol fuel alcohols appear promising alternatives to increasingly uncertain supplies of high-cost liquid fossil fuels in developing countries. These fuels can be produced from locally available feedstocks, which are often higher-yielding and cheaper to produce in developing nations than in typical industrial countries. A number of important policy questions need to be addressed, however, if these alcohol fuels are to be used to maximum advantage by developing countries. At present, costs for producing ethanol from sugar crops range from $500 to $800 per ton oil equivalent in 1975 dollars and anywhere from $400 to $1,000 from wood and other crops; costs of methanol may range from $200 to $300 (compared with the recent U.S. costs of gasoline from imported oil of $350 or so per ton oil equivalent—a figure that will of course, rise with any increase in oil prices). Lowering costs of ethanol feedstocks appears more promising for wood than for field crops.

While the mid-term cost prospects for methanol appear more attractive than those for ethanol, two key technical issues are not yet resolved for methanol: (1) the engineering feasibility of a wood-methanol plant at "reasonable" cost, and (2) the environmental effects, in particular the toxicity of methanol itself (see chapter 9). Developing countries could decide to leave development of methanol technology up to the industrial countries, but most methanol research in the United States and Germany involves coal and not biomass, so a wood-based technology may not be quickly forthcoming.

The appropriate village technologies are a mixed bag ranging from very small, efficiency-related improvements in present appliances using traditional fuels up to complex modern technologies like photovoltaic cells. It is impossible to make reliable cost calculations on the basis of the types of data reviewed here, but on a purely economic basis, these technologies do not appear to be superior to conventional alternatives, such as rural electrification.[69] But even so, the range of end-uses considered is wide, and some individual technologies out of the larger number available appear promising for some end-uses having particular technical needs. Small-scale hydroelectric plants should receive attention from developing-area planners, preferably as part of an overall revitalized hydroelectric program. Biogas deserves widespread testing and evaluation. Promotion of improved stoves for fuelwood and waste burning seems to be an effective and relatively low cost option for conserving energy (and forests) on the village level.

Individual village technologies, even when expensive by comparison with some larger scale sources of energy, could perhaps be justified on special grounds, such as providing better income distributions or greater employment in rural areas; these goals would fit in with those of integrated rural development schemes. Furthermore, the related value of village-oriented energy sources in fostering community co-operation, individual self-reliance, and other life-style changes could be considered of value. Whether these corresponding benefits outweigh the financial costs of many of the more expensive technologies is not a question that can be readily answered because so little is known both about producing such benefits and the values of such benefits to society. Here one can only bear in mind the usually correct tag of the economist that "there is no such thing as a free lunch" and that gaining these extra benefits will usually generate corresponding extra costs in using appropriate village energy technologies.

If any single theme runs through this treatment of two key sectors—electricity and liquid fuel replacement—and the separate question of alternative technologies for rural energy needs, it is the importance of one indirect solar energy resource: biomass. While fossil fuels like coal or new hydropower may yet save the day for electricity, many countries will apparently have only biomass to fall back on for both electricity and liquid fuels—if worse comes to worst. And biomass, with newer technologies like biogas from fermentation of wastes, will continue to be a key element in rural energy use. The need to understand the land-use tradeoffs involved in biomass for energy therefore becomes a high priority task for research.

[69] See, for example, Cecelski and coauthors, *Household Energy*, chapter III and table III-1.

APPENDIX 8-A
COMPARATIVE ETHANOL AND METHANOL COSTS

The tables show ethanol costs (table 8-A-1) and methanol costs (table 8-A-2) from various sources.

Note that the methanol cost uncertainties are of a somewhat different order than those of ethanol, since the process for making the latter is commercially established.

Table 8-A-1. Costs of Ethanol from Various Feedstocks
($1975 per gigajoule)[a]

Cost category	Poole[b]	Yang[c]		Lipinsky[d]	Katzen[e] [recalculated]		Scheller[f] (modified)		IGCE[g]
					Sources				
Capital charges	2.25	5.80	3.60	5.20	14.00 [4.80]		4.40		4.20
Feedstock — Type	Cane	Cane	Cassava	Cane	Wood waste (at 15/ODT)	(at 34/ODT)	Corn (2/bu.)	(3/bu.)	Wood (23/ODT)
Feedstock — Cost	8.65	9.60	10.80	10.80	3.80	8.60	8.40	12.60	5.70
By-products[h] (Credit)[i]	0.70	0.70	0.80	[j]	2.60		4.50	6.80	3.30
Other costs (O&M, labor, chemicals, etc.)	2.70	2.20	4.30	2.10	3.80		3.70		6.50
Unit cost (total)[k]	12.90	16.90	17.90	18.10	19.00 [9.00]	23.80 [14.60]	12.00	13.90	13.10

Note: ODT = oven-dry ton

[a] To convert $1975/per gigajoule into $1975/per metric ton oil equivalent, multiply the $1975/GJ by 43.1; for example, $1/GJ = $43.1/toe.

[b] Derived from Alan Poole with Jose Roberto Moreira, "A Working Paper on Ethanol and Methanol as Alternatives for Petroleum Substitution in Brazil," mimeo (São Paulo, Instituto de Física da Universidade de São Paulo, August 1979) p. 97.

[c] Derived from Victor Yang and Sergio C. Trindade, "The Brasilian Fuel Alcohol Program." Paper presented at the International Seminar on Energy. Hyderabad, India, January 1979, and reported in *Chemical Engineering Progress.* April 17, 1979, p. 17. Conversion factor is 21.2 GJ/m³.

[d] Derived from E. S. Lipinsky, *Sugar Crops as a Fuel Source.* vol. II. final report to the Energy Research and Development Administration (Columbus, Ohio, Battelle Columbus Laboratories, April 1978) p. 92. Assumes a 180-day season; costs are 33 percent higher for 90 days (p. 81). Conversion factor is 0.0803 GJ/gallon.

[e] Derived from Raphael Katzen Associates, *Chemicals from Wood Wastes* (Cincinnati, Ohio, December 1975) (available from the National Technical Information Service, PB 262-489). The values enclosed in square brackets are the results of a recalculation based on capital figures given in the Katzen reference on p. 87, with other parameters on pp. 86–89 of Katzen. In the recalculation, however, contingency is taken as 15 percent of direct costs, so that total capital costs are $64,000, and capital charges as 15 percent of that, or $9,600 per year. Conversion factor is 2 million GJ/25MM gallons.

[f] Derived from William A. Scheller, "The Use of Ethanol-Gasoline Mixtures for Automotive Fuel" (Orlando, Fla., Institute for Gas Technology, 1977), as modified by Thomas A. Sladek, "Ethanol Motor Fuel and Gasohol," *Mineral Industries Bulletin* vol. 21, no. 5 (May 1978) to reflect variable by-product credits. Conversion factor is 0.0803 GJ/gallon.

[g] Derived from Inter Group Consulting Economists, Ltd. (IGCE), *Liquid Fuels From Renewable Resources: Feasibility Study* (Winnipeg, Manitoba, March 1978). Capital and operating costs derived from Institute of Gas Technology (IGT), "Clean Fuels from Biomass and Wastes," symposium paper, Orlando, Fla., January 25–28, 1977, p. 17, with wood prices and by-product credits (for 1975) from p. 25, with $1977 (Canadian) taken as 1.2 $1975 (U.S.). Conversion factor is 26.9 GJ/metric ton.

[h] Not including bagasse, lignin, etc., by-products used as process fuels not included in other totals.

[i] These are credits for by-products that are subtracted from the costs.

[j] No separate credit for by-products taken.

[k] Compare U.S. mid-1979 gasoline wholesale costs of $3.50 ($1975/GJ) and gasoline from imported oil of about $8.00/GJ.

215

Table 8-A-2. Methanol Costs
($1975 per gigajoule)

Source	Feedstock and costs in original units	Costs
Poole[a]		
Poole recalculation	wood at $1.25–1.75/GJ	7.10–8.15
CESP calculation	wood at $1/GJ	6.50
Lipinsky[b]	synthesis gas	2.9 (plus gas cost)
ERDA[c]	wood { at $1/GJ	4.0–8.0[d]
	at $2/GJ	6.0–9.0[d]
ERT[e]	coal	2.2–3.3
Reed[f]	methane at 40¢/10³ft³	2.0
Katzen[g]		
Original Katzen calculation	wood { at $15/ODT	12
	at $34/ODT	14
Recalculated at 10% capital charges	wood { at $15/ODT	5.2
	at $34/ODT	8.4
Recalculated at 15% capital charges	wood { at $15/ODT	6.2
	at $34/ODT	9.3

Notes: GJ = 10^9 joules = 0.95 million Btu = .0232 toe. ODT = oven-dry ton, about 20 GJ.

[a] Derived from Alan Poole with J. R. Moreira, "A Working Paper on Ethanol and Methanol as Alternatives for Petroleum Substitution in Brazil" (São Paulo, Instituto de Física da Universidade de São Paulo, 1979) pp. 135–136.

[b] E. S. Lipinsky, S. Kresovich, and coauthors, "Systems Study of Fuel from Sugarcane, Sweet Sorghum, and Sugar Beets," third quarterly report to the Energy Research and Development Administration (Columbus, Ohio, Battelle Columbus Laboratories, April 14, 1976) p. 96.

[c] David L. Hagen, *Methanol: Its Synthesis, Use as a Fuel, Economies, and Hazards* (Washington, D.C., Energy Research and Development Administration, 1976) p. II-12.

[d] One gallon is taken here as 0.061 GJ.

[e] Energy Resources and Technology (ERT), "Westinghouse Engineers See Zero Emissions Power Plants Fueled by Methanol" (Silver Spring, Md., December 15, 1978) p. 493.

[f] T. B. Reed and R. M. Lerner, "Methanol: A Versatile Fuel for Immediate Use," *Science* vol. 182, no. 4119 (December 28, 1973) p. 1301.

[g] Raphael Katzen Associates, "Chemicals from Wood Wastes" (Cincinnati, Ohio, December 1975) p. 46 (available from National Technical Information Service, PB 262-489).

APPENDIX 8-B
LAND REQUIRED TO REPLACE LIQUID FUELS WITH ETHANOL AND METHANOL

The following tables in this appendix compare selected developing countries' arable and forest lands, possible alcohol production, and the relation of this theoretical production to 1976 liquid fuel consumption.

Note that alternative land uses are *not* considered in these theoretical calculations: in practice, a nation would be able to produce less alcohol—often far less—because of competing needs for land.

Table 8-B-1. Land Required for a Hypothetical Replacement of Liquid Fuels with Ethanol from Agricultural Crops

Feedstock & country	Average yields (metric tons/ha)	Alcohol production[a] (liters/ha)	Liquid fuel consumption[b] 1976 (mil. liters)	Land required to meet 1976 liquid fuel consumption[c] (1000 ha)	Total arable and permanent cropland, 1975 (1000 ha)	Total forest & woodland, 1975 (1000 ha)	Total arable & forest land, 1975 (1000 ha)	Percentage of "available" land required to meet 1976 liquid fuel consumption[d]
Sugarcane								
Brazil	46	3,035	54,000	27,000	36,600	510,000	546,600	5–74
Cuba	45	2,970	9,900	5,000	3,110	1,240	4,350	110–160
Dom. Rep.	64	4,225	2,800	1,000	995	1,104	2,099	48–100
USA	85	5,610	980,000	260,000	209,236	304,400	513,636	51–120
Egypt	79	5,215	13,000	3,700	2,862	2	2,864	130
India	51	3,365	25,000	11,000	167,200	67,400	234,600	5–7
Indonesia	84	5,545	22,000	5,900	18,600	121,400	140,000	4–32
Philippines	49	3,235	12,000	5,500	7,899	12,300	20,199	27–70
Sweet sorghum[e]								
Ethiopia	52	4,044	620	230	13,730	8,860	22,590	1–2
Nigeria	52	4,044	4,000	1,500	23,750	31,069	54,819	3–6
Sudan	52	4,044	2,200	810	7,495	91,500	98,995	1–11
Upper Volta	52	4,044	100	37	5,613	3,675	9,288	0.4–0.6
India	52	4,044	25,000	9,200	167,200	67,400	234,600	4–6
Argentina	52	4,044	9,900	3,700	34,550	60,700	95,250	4–11
USA	52	4,044	980,000	360,000	209,236	304,400	513,636	70–170
Corn								
Kenya	1	340	1,700	7,300	1,765	1,935	3,700	200–420
Malawi	1	340	160	720	2,278	2,314	4,592	16–32

Tanzania	1	340	870	3,800	6,070	31,074	37,144	10–63
USA	5	1,700	980,000	860,000	209,236	304,400	513,636	170–410
El Salvador	2	680	850	1,900	651	250	901	210–290
Argentina	2	680	9,900	22,000	34,550	60,700	95,250	23–63
Turkey	2	680	18,000	40,000	28,286	20,170	48,456	83–140
Thailand	2	680	11,000	24,000	16,580	20,500	37,080	64–140
Cassava								
Cameroon	4	696	420	900	7,345	30,000	37,345	2–12
Ghana	9	1,566	970	920	2,700	2,447	5,147	18–34
Nigeria	10	1,740	4,000	3,400	23,750	31,069	54,819	6–15
Indonesia	8	1,392	22,000	24,000	18,600	121,400	140,000	17–130
Sri Lanka	5	870	1,100	1,900	1,979	2,899	4,878	40–96
Thailand	18	3,132	11,000	5,200	16,580	20,500	37,080	14–31
Brazil	13	2,262	54,000	36,000	36,600	510,000	546,600	7–98

Sources: *Average yields, total arable & permanent cropland,* and *total forest and woodland* from United Nations Food and Agriculture Organization, *1976 Production Yearbook,* vol. 30 (Rome, 1977). Liquid fuel consumption from United Nations, *World Energy Supplies, 1972–1976,* Series J, No. 21 (New York, 1979).

a Alcohol production per ton of feedstocks based on reported current yields as follows: *sugarcane* (fresh stalks) 66 liters/mt, from J. Goldemberg, "Brazil: Energy Options and Current Outlook," *Science* vol. 200, no. 4338 (1978) p. 163. *Sweet sorghum* (fresh stalks) 78 liters/mt (sweet sorghum is not presently widely produced commercially); yields are based on E. S. Lipinsky, *Sugar Crops as a Fuel Source,* vol. II, final report to the Energy Research and Development Administration (Columbus, Ohio, Battelle Columbus Laboratories, April 1978); projected average 1980 yields of 52 mt/ha in southern United States, 6.8 mt/ha fermentable sugars, p. 25; assuming 50 percent conversion into ethanol, yields 3.2 mt/ha ethanol = 4,044 liters/ha. *Corn* (grain) 340 liters/mt, from Stanford Research Institute International, *Mission Analysis for the Federal Funds from Biomass Program* (Washington, D.C., U.S. Department of Energy. December 1978) p. 103 (available from the National Technical Information Service). *Cassava* (fresh) 174 liters/mt, from J. Goldemberg, "Brazil," p. 163.

b 1.5 mtce = 1 mtoe = 1 million tons oil equivalent.

c Assuming 1.33 liters ethanol/liter of liquid fuels and that land of similar productivity is available and utilized.

d Lower percentage is of total arable, permanent crop, forest, and woodlands; higher figure is of only currently arable and permanent crop lands.

e Sweet sorghum has not been widely produced commercially; yields are assumed constant (see footnote a to this table).

for example, Brazil's 1976 fuel consumption is 54,000 million liters divided by 1,333, which approximately equals 40.5 million tons

Table 8-B-2. Land Required for a Hypothetical Replacement of Liquid Fuels with Ethanol or Methanol from Forest Crops

| Alcohol | Liquid fuel consumption 1976 (million liters)[c] | Total forest & woodlands 1975 (1,000 ha) | Forest land to meet 1976 liquid fuel consumption at yields of: | | | | | |
| | | | 4 t/ha[a] | | 8 t/ha[b] | | 20 t/ha[b] | |
			(1,000 ha[d])	(% total)[c]	(1,000 ha[d])	(% total)[c]	(1,000 ha[d])	(% total)[c]
Ethanol								
USA	980,000	304,400	2,096,501	689	1,048,250	344	419,300	138
Brazil	54,000	510,000	115,521	23	57,761	11	23,104	5
Columbia	8,200	77,190	17,542	23	8,771	11	3,508	5
Kenya	1,700	1,935	3,637	187	1,818	94	727	38
Ethiopia	620	8,860	1,326	15	663	7	265	3
Bangladesh	1,000	2,212	2,139	97	1,070	48	428	19
India	25,000	67,400	53,482	79	26,741	40	10,696	16
Indonesia	22,000	121,400	47,064	39	23,532	19	9,412	8
Methanol								
USA	980,000	304,400	1,164,587	382	582,294	191	232,917	77
Brazil	54,000	510,000	64,171	13	32,085	6	12,834	3
Columbia	8,200	77,190	9,745	13	4,872	6	1,949	3
Kenya	1,700	1,935	2,020	104	1,010	52	404	21
Ethiopia	620	8,860	737	8	368	4	147	2
Bangladesh	1,000	2,212	1,188	54	594	27	238	11
India	25,000	67,400	29,709	44	14,854	22	5,942	9
Indonesia	22,000	121,400	26,144	22	13,072	11	5,229	4

Sources: Liquid fuel consumption from United Nations Food and Agriculture Organization, 1976 Production Yearbook, vol. 30 (Rome, 1977). Total forest & woodlands from United Nations, World Energy Supplies 1972–1976, Series J, no. 21 (New York, 1978).

[a] A yield of 4 t/ha (1.4 toe) is chosen as within the range of the yields of 1–10 m³ (0.23–2.3 toe) assumed in chapter 7 for existing forests. One t of air-dried wood here is assumed to be 0.355 toe and also about 0.765 ODT (oven-dried ton); see note[d] below.

[b] Estimated future yields for temperate forests: 8 t/ha, for an expected whole-tree potential under average current conditions; 20 t/ha, if a program of fertilization and genetic improvement are included. See William F. Hyde and Frederick J. Wells, "The Potential Energy Productivity of U.S. Forests," mimeo (Washington, D.C., Resources for the Future, 1975) table 4, p. 21.

[c] 1.5 toe = 1 toe – 1,333 liters; see footnote b in table 8-B-1 for an example of conversion of units.

[d] Alcohol production per ton of wood based on current reported yields: 1 ODT (oven-dried ton, about 0.46 toe) wood = 228 liters ethanol (average of 200.4–255.6, depending on species); 1 ODT aspen chips, 12 percent bark = 550 liters methanol. Assumes 1.33 liter ethanol replaces one liter of liquid fuels and 2 liters methanol replaces one liter of liquid fuels and that land of similar productivity is available and utilized (Inter Group Consulting Economists, Ltd., "Liquid Fuels from Renewable Resources: Feasibility Study" [Winnipeg, Manitoba, March 1978]) pp. 19, 73.

[e] Hypothetical percentage of forest land that would have to be utilized to replace 1976 liquid fuel consumption with alcohol.

9

Supply Planning Problems

Finding resources and technologies to produce energy at reasonable costs is of course only the first step in achieving any improvement in energy supplies. There are all kinds of economic constraints besides costs. Bottlenecks like labor shortages and credit availability can be important. Infrastructure can be missing, market signals can go awry, and other national goals like foreign exchange and employment objectives are all crucial in planning decisions. The environmental and health impacts of energy technologies have been playing a crucial role in industrial countries. They are also of growing concern to planners in developing areas. Furthermore, the socioeconomic context is important. Energy is not a good in itself, but only contributes to other goods and services. Mere energy balances in tons of oil equivalent are not adequate as guides for realistic planning: there are always individual preferences within the total socioeconomic structure of any society that have to be taken into account. Finally, equity problems are a real concern in economic development, and the energy supply problem is no exception.

Economic Constraints

Apparent costs of energy technologies may not give the whole picture, even if adapted to specific developing areas. Important economic constraints other than conventional cost factors can be present. While some of these have been considered briefly in chapter 8, other basic analytic problems of interest to the planner are considered here:

1. The bottleneck problem can be important. The assumption often made in cost–benefit analysis that factors of production—

capital, labor, land, and management—are available at a given price in unlimited quantities may be wrong. Supply curves can be complex, and some factors may not be available at (apparently) reasonable costs. In fact, full-fledged "lumpiness" (or indivisibility) problems could arise; for example, credit may not be locally available at any price. Needs for external financing can also be a complex fact of life.

2. Existing deficiencies in infrastructure, for example, in transportation, could greatly increase costs of some new energy systems. Other development objectives would have to be combined with energy needs in some cases to justify suitable modifications to infrastructure.

3. The energy crisis is a problem of national as well as individual concern. But economic signals to the individual firm or public agency may be at variance with national economic interests.

4. The cost of an energy technology—even when corrected for the market failure and social accounting problems just mentioned—will not reflect its impact on other social goals such as securing full employment.

Bottlenecks in Obtaining Inputs for Energy Systems

Needed economic factors. Specialized labor and materials may be required for building new types of energy systems, such as photovoltaic generators, with concomitant difficulties in estimating the opportunity costs of such factors. For one thing, supplies of specialized land, labor, and capital items may be locally in very short supply—or not available at all. The supply curve in such cases will be highly inelastic: costs may rise very sharply as more energy projects are implemented. This means that in evaluating a new energy project of even modest size, one may have to take into account other energy projects and indeed general development activities that might compete for resource use.

For example, even though "soft" technologies have been praised as inherently more participatory than some alternatives,[1] solar energy in the U.S. context, at least, still makes use of fairly complex kinds of labor inputs. According to a recent Battelle-Columbus study,[2] the

[1] Amory B. Lovins, *Soft Energy Paths: Toward a Durable Peace* (Cambridge, Mass., Ballinger Publishing Company, 1977) p. 151.

[2] Quoted in *Solar Energy Intelligence Report* (Silver Spring, Md., Business Publishers, January 22, 1979) p. 36.

labor forces of U.S. commercial solar companies included only 29 percent unskilled workers (with 30 percent skilled, and the rest professionals), while in a sample of installers, 41 percent were unskilled (with 39 percent technicians and skilled labor and the rest professionals).

A difficult and related problem involves "lumpiness," or indivisibility, of inputs. It may be economically feasible to produce some energy outputs or technologies only in large quantities. In fact, certain types of intermediate goods, such as parts and components for small-scale hydroelectric or large-scale wind generators, might be essentially unavailable at low volumes of demand—even as imports, much less as locally produced goods. Or it could become virtually impossible to obtain small nuclear plants from reactor manufacturers now geared to 1,000 megawatt (Mw) facilities in industrial countries.

Various technologies would experience these bottlenecks in differing ways. Photovoltaic units might be difficult to buy, but easy to maintain—providing the difficulties mentioned in chapter 8 are overcome. Although the attractiveness of some biogas schemes is greatly enhanced by the use of fairly simple structural materials and building designs that do not call for imported technology, even for this technology some of the labor constraints can be important: in particular, training of operators and repair personnel has been noted as a major obstacle to wider use.[3] As already discussed, especially in chapter 8, land costs for biomass systems may become high or even infinite; that is, land might be locally unavailable, depending on alternative uses and land tenure arrangements.

Although such supply anomalies and market failures may be severe, in many cases they can be circumvented by factor substitution—for example, labor-intensive alternatives for imported capital items—or by government-aided import arrangements. In practice, the two greatest defects in markets for the supply of new and unfamiliar energy technologies may be lack of *information* and of *credit*.

Information about new sources, sources of suppliers, and requirements for labor skills may all be missing in some local context. Indeed, it is evidently assumed in many project analyses that essential information, and sometimes management skills, will be supplied by sources external to the project area. But information and management skills are often not available in practice in many developing-area situations.

[3] Elizabeth Cecelski, Joy Dunkerley, and William Ramsay, *Household Energy and the Poor in the Third World* (Washington, D.C., Resources for the Future, 1979) p. 101.

A special bottleneck: local credit constraints. A key economic constraint on supply is the availability and cost of capital. The opportunity cost of capital is a central question for many of the capital-intensive energy sources. Thus, a vital question particularly difficult to answer in the context of a developing area can be: What is the appropriate discount rate (the true cost of lending money)? In Latin America, for example, interest rates from private lenders are often very much higher than those from public agencies. The analysis is complicated by inflation, where interest rates have to become artificially high just to make up for shrinking currency values. But if one corrects for that, government interest rates ranged from − 16 to 8 percent on a deflated basis in one review of government credit organizations in several countries, compared with rates (on a deflated basis) of over 20 percent for private lending.[4] The relationship between private discount rates and "social" (government) discount rates is involved here; that is, an average private investor may legitimately value his loan money more highly than the public as a whole acting through government agencies lending public funds. Different perceptions of risks and uncertainties are especially relevant; because, for many government investments, risk premiums for loans may be greatly understated in practice. Correcting for this understatement would probably draw the private and social rates much closer together; indeed the proper rate for many government-sponsored energy projects should probably be some compromise between the two rates.[5] But it must be recognized that this question of correct interest rates is complex and that analysis in this area is still evolving.[6]

A practical aspect of concerns about the opportunity cost of energy capital investment is that special loans may be available for many economic development purposes at low interest rates, and energy projects may find themselves in competition for these low-cost loans.

[4] World Bank, "Agricultural Credit: Sector Policy Paper" (Washington, 1975) p. 79.
[5] Many support the idea of special social rates (Amartya K. Sen, "Isolation, Assurance and the Social Rate of Discount," *The Quarterly Journal of Economics* vol. 81, no. 1 [February 1967] pp. 112–124). Others (William J. Baumol, "On the Social Rate of Discount," *The American Economic Review* vol. 58 [1968] p. 796) believe that the private rate is the proper choice. The latter view, however, appears to depend on the fact that the private investment is taken as the "normal" case, and its marginal costs are taken as the marginal costs for all capital. This may be very misleading in economies in which government investment is the most important component.
[6] See, for example, World Bank, "Social Cost-Benefit Analysis: A Guide for Country and Project Economists to the Derivation and Application of Economic and Social Accounting Prices," Staff Working Paper No. 239 (Washington, D.C., 1976) pp. 38–42.

This means that the prevailing rate for such loans could turn out to be the relevant opportunity cost of capital. If so, the apparent financial cost of capital-intensive technologies like hydroelectricity, active solar schemes, and nuclear power would be greatly reduced. The real cost, however, would be higher, but part of it would be borne by outsiders (in the case of external loans) or by the public at large (in the case of subsidized domestic loans).

Some energy projects, on the other hand, may be financed from private sources. In that case, not only will the relevant interest rate usually be higher, but the question of capital availability itself arises. Many countries have been slow to develop well-organized capital markets.[7] Moreover, in many nations, there may be no real shortage of domestic savings, but domestic capital markets are unable to bring savers together with borrowers waiting to undertake potentially high-yield projects. Strong oligopolistic (market-restrictive) overtones were observed in a study of banking in one developing country, with wide gaps between interest paid on deposits and that charged on loans.[8] Even when banking is fairly competitive, bankers may view risks involving new types of energy projects as prohibitively large, therefore making private capital essentially unavailable. However, innovative banking procedures, such as risk assessment at a local level, door-to-door deposit collection, and especially, decentralized decision making can help to expand credit opportunities, as in the case of the Syndicate Bank in India.[9]

Government energy agencies can have a direct influence. For example, the actual costs of credit extended under the aegis of the Khadi and Village Industries Commission for domestic biogas plants in India may have actually risen under recent changes in the law, since out-and-out subsidies have been reduced or eliminated.[10] However, the numbers and locations of credit outlets have been greatly increased and easier credit availability may have in fact led to recent spurts in rates of digester construction in India.[11]

[7] William G. Demas, *The Economics of Development in Small Countries with Special Reference to the Caribbean* (Montreal, McGill University Press, 1965) p. 30.

[8] Andres Ostlind, *Development Policy and Credit Institutions,* Fiftieth Anniversary Commemoration Lectures (Cairo, National Bank of Egypt, 1969).

[9] World Bank, *Capital Market Imperfections and Economic Development,* Staff Working Paper No. 338 (Washington, D.C., 1979) pp. 13–25.

[10] H. R. Srinivasan of the Khadi and Village Industries Commission, Bombay, to William Ramsay, January 1979.

[11] World Bank, *Capital Market Imperfections,* p. 86.

In practice, informal credit arrangements may be of great importance. Agricultural traders can supply credit services, or richer farmers can lend to the poorer. In one case, a lowering of interest rates was observed as more knowledgeable lenders entered the credit market and risk premiums therefore dropped.[12] The availability of foreign capital is another issue, which melds into the complex question of balance of payments (see chapter 4 above).

External financing of new energy supplies. In addition to local credit constraints, there is also the question of whether national financing as a whole will be adequate to finance new energy supplies, or whether funds from other countries or international entities will be forthcoming in the requisite amounts. In a recent study, Gordian Associates have made estimates for the development of energy sources in the non-OPEC developing countries over the next forty years or so (for the special case where energy self-sufficiency is assumed to be the goal) based on projections of the amounts of capacity needed in developing areas to produce new energy from oil, coal, gas, hydro-electric, and nuclear power. The results are about $390 billion for the case where real energy prices increase, and about $870 billion (in $1976) for the constant real energy price case, the difference reflecting conservation effects induced by rising prices.[13]

How much of those funds will actually be available? The conclusions of the Gordian study were that, despite past difficulties, the financing for oil and gas exploration should be easy to come by in the future. At the same time, that study was much less optimistic about coal, because it postulated that coal trading would not be profitable and was pessimistic about the interest of international private equity in national coal programs. Furthermore, despite the interest of such institutions as the World Bank in hydroelectric power, the fact that projects are often not self-liquidating was viewed as a serious drawback to attracting outside investment. In sum, the study predicted that perhaps 20 percent of financing needs in either case would be a "financing shortfall."[14]

[12] Ibid., pp. 7–10.

[13] Gordian Associates, *Energy Supply and Demand Balances and Financing Requirements in Non-OPEC Developing Nations* (Washington, D.C., U.S. Department of Energy, 1979) p. 77 (available from the National Technical Information Service).

[14] The incremental oil import burden needed to replace this "inadequate finance capital" would be some 84.5 billion dollars in the year 2020 for increasing real energy prices, compared with 77.3 billion dollars in capital shortfall (Gordian Associates, *Energy Supply and Demand Balances,* pp. 87–88). Therefore the net annual burden from not having the financing is of the same order of magnitude as the missing financing.

These estimates, however, were made for a hypothetical energy self-sufficiency scenario. Furthermore, any movement on the part of developing nations toward energy self-sufficiency should act to lower the international price of oil. It also seems generally agreed that the potential for renewable resources is large enough to cause a significant easing of overall supply constraints.

Infrastructure Problems

The general economic infrastructure may also be the source of market-related elements that can both raise costs of new energy systems and lower the cost of older competing systems. Not all such effects can be easily traced: strictly speaking, the structure of the economy would have to be examined in detail in order to determine the microeconomics of any particular renewable energy project. But particular cases may be of interest because something can be done about them; prime examples are such infrastructure elements as highway transportation.

Transport costs are critical for some biomass projects, especially for large biomass-fired electricity, gasification, or liquid production installations. For example, the presence or absence of a suitable road network can be the critical element in determining the costs of supplying fuel as a function of distance. Moreover, fuel costs will be greater in general for larger plants, because fuel must come from longer distances. But one can expect larger plants to be cheaper per unit of capacity than smaller plants. The tradeoff of infrastructure and plant costs therefore becomes of prime interest in the feasibility analysis. Infrastructure itself is often lumpy by nature: after all, one cannot usually purchase marginal amounts of roadway. On the other hand, roads are built for purposes other than energy supply, and energy schemes can piggyback on other development projects.

A contrasting case involving infrastructure constraints is that of centralized electric grids. If such grids are already available in any area, they represent to a large extent a sunk cost, and as such, a fact of life in determining energy price alternatives to potential decentralized sources. A newly proposed electric grid, however, could constitute a more complex problem. Transmission lines are "extensive" by nature, and a proposed solar energy project in one area might have to contend with the fact that a competing new electric grid could be justified for its benefits to other areas outside the neighborhood of the solar project.

Indeed, a local decentralized project could turn out retrospectively to have incurred much higher costs than originally calculated if a central grid must later be built to service the same geographical area.

Therefore, although infrastructure-related energy supply costs could be regarded as fixed in the usual static analysis, national planners may properly wish to consider new energy options as part of a complex of other needs for agriculture and industry. Implementing a complex of new projects would then require reassessing local infrastructure needs, and any resulting infrastructure changes would then affect the final calculation of energy project costs.

National Versus Individual Accounting

To an individual investor in an energy facility, costs may appear to be different than they do to a national policy maker. In particular, the role of uncertainty can be very important in making judgments about what costs for new sources are acceptable relative to existing energy sources. For example, the private cost of renewables—properly corrected for the constraints just reviewed—would then have to be less than those of competing fossil fuels, if private individuals or firms were to undertake new renewables projects. Of course, such private investors should consider that even if oil prices were to appear relatively low at present, they may rise to much higher levels in the future—even though there are good reasons for thinking that possible price rises may ultimately be capped at levels twice 1978 prices by costs of future synthetic fuels.[15] However, even if such a price rise appeared probable, the possibility that it would not occur might pose too much of a risk for a private investor. On the other hand, national policy could encourage new sources *despite* oil price uncertainties because on a national level relatively modest risks of misallocating resources could be accepted. This discrepancy between individual and national accounting outlooks could prevent the successful implementation of energy plans unless corrected by some kind of loan guarantee, insurance scheme, or pricing agreement.

Other discrepancies of a similar sort can arise: in particular, there may be social reasons for subsidizing relatively expensive renewable sources to replace fossil fuels. Prime examples are the well-known environmental and health externalities (see below). Governments could

[15] Sam H. Schurr, Joel Darmstadter, Harry Perry, William Ramsay, and Milton Russell, *Energy in America's Future: The Choices Before Us* (Baltimore, Johns Hopkins University Press for Resources for the Future, 1979) p. 29.

give cleaner energy sources special encouragement to avoid problems such as air pollution and oil spills. An even more important source of discrepancy between the two points of view arises from the national policy maker's concern with protecting the country from sudden cutoffs of imported oil.

Another difference in private and social accounting systems occurs with traditional fuels. Although not all fuelwood is "free," in many rural areas it is gathered without monetary cost by members of the community—sometimes women and children—who might otherwise have little economically productive use for their time. However, the hidden costs in this situation may be large. For example, children could theoretically be educated in the time now spent gathering fuelwood for other, more productive tasks. The cost of this "free" activity to the wider community can be substantial for other reasons too. Overuse of wood can result in deforestation, changing drainage patterns, and reduced agricultural productivity, leading ultimately to the need to import food. In this case, what is free to the private individual is clearly very costly to the community at large. But because woodgathering is seen as free to the consumers, they have no incentive to use less energy by installing more efficient wood burning stoves; on the contrary there is a strong disincentive, as even the cheapest of these stoves would cost more than the existing appliance on a private accounting basis. Differences in social and private costs of this nature provide a strong basis for government intervention to compensate for such disparities. In the case of the "free" use of fuelwood, this might take the form of distributing improved stoves without charge.

Finally, in rural areas of many regions, the village is virtually a self-sufficient energy unit, producing and consuming energy within its own territorial limits. "Imports" from outside the area are negligible, as are "exports." In these circumstances, changes in local energy-consuming habits tend to depend on counterpart changes in local energy-producing systems. This is not always easy because the energy supply system is closely integrated with the existing distribution of wealth in the village and particularly with patterns of land holding. Any proposed change means that there will be both gainers and losers within a village, and therefore always some opposition to the proposal.

Satisfying Other National Goals

Cost-effectiveness criteria, even if corrected for complications like indivisibilities and other market failures, infrastructure problems, and differences between national and individual accounting, may still not be an adequate measure of the desirability of an energy option.

Equity problems are discussed later in this chapter and other national goals, such as balance-of-payments improvements, are discussed elsewhere in this book (see, for example, chapter 5).

Environmental Impacts

The effects that new and expanded energy technologies may have on the environment in developing areas commonly receive relatively little attention. The one big exception is deforestation, as noted in chapter 2. Several reasons probably account for this neglect. First, the commercial energy sector is apt to be a rather small part of the total economy, so that any health and environmental impacts may currently be at low levels relative to industrial-country experience. Second, commercial energy, by and large, tends to be disproportionately concentrated in oil, and the impacts from electricity generation by oil are typically much less than from many potential substitutes—including coal and nuclear power and also some biomass options. Hydroelectric dams certainly have large environmental impacts, but do not involve air pollution, radiation, or the production of carbon dioxide. The third reason for environmental neglect can be simply an embarrassment of environmental riches[16] in that even though ratios of population to good farmland are often high, some developing countries have no overall shortage of unspoiled (and unutilized) land and other environmental "goods." But, it is also true that other countries do not have large amounts of unspoiled natural resources, and some kinds of environmental problems are beginning to loom large. A fourth reason for environmental neglect is the very high priority that planners often give to economic development imperatives.

As development proceeds under new energy regimes, the first three motives for assigning a relatively low priority to environmental impacts will become less important. The fourth reason—the development imperative—may or may not. But even if development maintains a general priority over environment in planning priorities, it is important to ensure that a correct balance of overall national goals—economic and environmental—is maintained in the future. How this correct

[16] Compare William Ramsay and C. Anderson, *Managing the Environment, An Economic Primer* (New York, Basic Books, 1972) pp. 113–114.

balance is to be defined is ultimately a political matter. But information will be needed for any type of environmental policy decision. Therefore, monitoring procedures for environmental impacts should be established early on so that national policy making can benefit from adequate environmental information.

Admittedly, the issues of environment (including health) involve a complex, nonlinear system of causes and effects. Some energy-related projects, such as energizing pumpsets, could lead to net increases in environmental quality through the provision of potable water. And, in general, any economic development can be expected to lead to increased goods and services and therefore greater public well-being— as well as to a series of inevitable impacts on the local environment. Furthermore, projects in the developing countries will undoubtedly begin to fall into the common pattern now observed in many industrial countries: a development project will be allowed to proceed, but its execution will be accompanied by more or less stringent measures to abate environmental damage.

None of this, however, obviates the need to assess the direct effects of increasing different types of energy supplies and technologies. While environmental impacts are myriad in number and wide in scope, there are several key potential environmental problems with any technology, specifically with petroleum, electricity sources in general, alcohol fuels, and appropriate village technologies.

A general increase in use of petroleum products can be expected to have harmful consequences for health. With the exception of oil-producing countries, which will have to deal with the rather high frequency of accidents in the oil industry,[17] the single most important oil health problem may be associated with air pollution. For stationary sources, this problem is not as bad as it might first appear because effective methods exist for scrubbing exhaust gases from oil-fired power plants to remove sulfur and particulates and because strong scientific doubts have been expressed in recent years about early predictions of a high level of risk from these emissions.[18] But, for motor vehicle emissions, the dangers to health, while also subject to some uncertainty, must be taken extremely seriously. This is especially true

[17] U.S. Environmental Protection Agency, *A Summary of Accidents Related to Non-Nuclear Energy* (Washington, D.C., 1977) pp. 3–4 (available from the National Technical Information Service).

[18] See William Ramsay, *Unpaid Costs of Electrical Energy* (Baltimore, Johns Hopkins University Press for Resources for the Future, 1979) chapter 2, for a brief review and critique of the literature.

for the production of photochemical smog from vehicles. Pollution control for internal combustion engines and for all sources of unburned hydrocarbons and nitrogen oxides will therefore require attention by regulators in many developing countries. Another concern, the environmental effects of oil spills, is still a matter of some controversy.[19] It is possible that the long-term ecological damage from oil spills has not been as great as feared by many at the time of the first large spills, although short-term effects on plant and animal life are certainly significant.

Oil is one prime fuel for electricity; but other electrical power sources also have important environmental effects. If hydroelectric power is greatly extended, effects on local ecosystems are inevitable. When a particular nation contains large areas of wilderness, energy planners could choose to downgrade the importance of such impacts. But such an attitude may not be feasible in specific situations; for example, when ecological changes related to hydroelectric projects affect siltation patterns and hence downstream agriculture. Although water projects have been responsible for increases in parasitic diseases, the most dramatic effects on health are usually the result of the sudden failure of a dam.[20] While, for example, only one failure in about 1,000 hydroelectric dams has occurred in the United States, the Teton Dam accident in 1977 did cause eleven deaths and $1 billion in damages. Other failures (in both hydro and nonhydro dams) in various countries have resulted in large losses of life, often measured in hundreds of fatalities.[21]

When considering newer technologies, the situation becomes even more difficult to analyze because experience is still so limited. The qualitative environmental implications of nuclear power have already been discussed (in chapter 8) as they relate to the future outlook for this technology. In the present state of uncertainty about nuclear safety, it is difficult to make quantitative statements. It appears that if the safety hazards turn out to be overstated, nuclear power impacts will be significantly lower than those from coal, ignoring the complex problem of nuclear proliferation.[22] On the other hand, if pessimistic estimates of nuclear safety turn out to be correct, risks from nuclear plants could equal or surpass those from an equivalent coal-fired plant.

[19] U.S. Environmental Protection Agency, *A Summary of Accidents*, p. 4.

[20] Ibid., p. 7.

[21] Herbert Inhaber, *Risk of Energy Production*, AECB-1119/Rev.-2 (Ottawa, Atomic Energy Control Board, 1978) table K-6.

[22] Schurr and coauthors, *Energy in America's Future*, p. 367.

Impacts from alcohol fuels are difficult to specify because of lack of large-scale experience with their use. Nevertheless, certain analogies with conventional fuels can be exploited for some cases. For alcohol that has been produced from biomass, studies have been made of pollutants that result from its use in engine combustion, with tests so far showing varying results, depending on carburetor setting and other engine parameters:[23] some pollutants would probably be increased and others decreased by the use of alcohol as a motor fuel. Lower carbon monoxide emissions and increased amounts of nitrogen oxide are likely results, but the evidence is conflicting and shows no clear trend yet. However, alcohol combustion would also produce noxious aldehydes; the importance of this new impact can not yet be assessed.

But the greatest difficulty could turn out to be high toxicity. Pure ethanol does not raise toxicity problems in practical terms but methanol does. It has long been recognized as a chemical that can cause death or permanent injury after a short exposure to small quantities.[24] Methanol can be absorbed through the skin and by inhalation, and occupational limits are now set at a low—compared with many common chemicals—level of 200 parts per million (of air) per 8 hour work day.[25] Other studies, however, note that gasoline is also toxic; but methanol is particularly noxious because it can cause blindness. At any rate, the entire toxicity question requires study; in particular, modeling of ingestion pathways in the context of a methanol motor fuel economy is sorely needed.

The production of alcohol from biomass, or the use of biomass in any context, involves difficult problems of economic competition with other resources and general environmental impacts. These are generally lumped together under the heading of "land use." The economic land use consequences of using farmland for fuel growing have been discussed in chapter 8, but the general environmental consequences can be formidable even if presently unproductive forest land is used. If biomass is used as feed to supply electric generating plants, for ex-

[23] Kenneth R. Stamper, "50,000 Mile Methanol/Gasoline Blend Fleet Study—A Progress Report," and Aaron Svahn, "Methanol/Gasoline Mixtures in Four Stroke Auto Engine." Papers presented at the Third International Symposium on Alcohol Fuels Technology, Asilomar, Calif., May 28–31, 1979.

[24] American Petroleum Institute, *Alcohols: A Technical Assessment of Their Application as Fuels*, Publication No. 4261 (Washington, D.C., 1979) p. 25.

[25] David LeRoy Hagen, *Methanol: Its Synthesis, Use as a Fuel, Economics, and Hazards* (Washington, D.C., *Energy Research and Development Administration*, 1976) p. IV-2: Alan Poole and José Roberto Moreira, "A Working Paper on Ethanol and Methanol as Alternatives for Petroleum Substitution in Brazil," Mimeo. (São Paulo, Instituto de Física de Universidade de São Paulo, 1979).

ample, the amount of land involved can be very large, over ten times the number of hectares that may be stripmined for a corresponding coal plant.[26] The ecological consequences of such large fuel-forest management schemes could be exceedingly complex, involving species endangerment, watershed changes, and even regional climate modification.

For the appropriate village technologies, the environmental picture is somewhat different. Some of the important virtues of solar and wind energy and biogas are that they have very little adverse effect on health and the environment. Indeed, effects may appear to be trivial: in the industrial countries, one of the most frequently cited environmental effects of large wind generators is their possible interruption of television transmission. It is possible to associate health effects with various forms of solar collectors, for example, direct accidents such as roof falls and indirect effects such as industrial hazards in manufacturing components,[27] However, it is unlikely that health effects from such technologies will in fact turn out to be a major stumbling block.[28] In fact, the implementation of a technology like biogas could have positive health effects: the gases resulting from the burning of methane-rich biogas are much cleaner than the particulates emitted in the burning of traditional fuels such as wood and wastes. While the degree to which the smoke from such traditional fires is deleterious to health is not well determined, technologies like biogas—or solar cookers—can nevertheless be credited with a measure of clean-air environmental benefits. For biogas from human wastes, the energy technology can also serve an important sanitation function. And, in a more indirect effect, decreases in wood uses in those cases where forests are unmanaged or managed poorly would tend to decrease deforestation and erosion problems.

It follows that, despite any cost and institutional problems, village technologies could well contribute important benefits by alleviating—directly or indirectly, as they replace fossil fuels—the health and environmental impacts from energy technologies.

Socioeconomic Context

Energy technologies and resources are not ends in themselves, but are tools used in a social context to satisfy human needs. In considering the technologies by themselves, planners often look at them in isolation

[26] Schurr and coauthors, *Energy in America's Future*, pp. 371, 375.
[27] Inhaber, *Risk of Energy*, pp. G-1–G-10.
[28] Schurr and coauthors, *Energy in America's Future*, pp. 366n, 368.

from the social situation and try to determine the least expensive supply of a given quantity of energy to fulfill certain needs. But the total function of the energy source over time must be considered in the planning calculations. For example, one of the simplest kinds of social context problems can show up in such questions as reliability: the energy from a new technology may not be produced reliably enough so that it can be depended upon. This idea is not new: reliability is usually and correctly considered in any overall cost engineering calculation. But we have seen in chapter 8 that maintenance of new systems, in particular for unfamiliar renewable sources, can be a severe problem, and the importance of a total social systems engineering approach to these problems must be emphasized.

A more fundamental and troublesome attribute of many renewable technologies is that for some systems, a certain amount of energy unavailability is more or less predictable. The sun does not shine at night, and relatively little energy comes through on a cloudy day. In contrast to industrial countries where the importance of having storage and backup for solar systems is well recognized, for developing areas there is sometimes an assumption that backup is costless (by reverting to traditional wood fires, for example) or that potential energy users should adjust to the operating cycle of the energy source rather than vice versa. This last assumption about human adaptability is evidently often not well founded. For example, there is evidence that rural villagers in parts of Africa object to cooking outside with solar cookers and to being unable to cook at night.[29] And such problems are not restricted to the rural sector. The successful use of alcohol fuels will also depend to a great extent on the ability of individual motorists to cope with and accept driveability difficulties like cold start problems.

Difficulties may also arise from energy planning that fails to take into account either amenities of life that are provided by traditional usages or other uses of a particular technology. Replacement of traditional wood or waste fires by other means, whether kerosine stoves or solar cookers, will not give the food that sometimes prized smoky taste; furthermore, functional energy analysis may not take into account other purposes satisfied by primitive fires, for example, incidental water heating and the psychic amenities of social interactions around a campfire.

[29] J. D. Walton, A. H. Roy, and S. H. Bomar, Jr., *A State-of-the-Art Survey of Solar Powered Irrigation Pumps, Solar Cookers, and Wood-Burning Stoves for Use in Sub-Sahara Africa,* report to Al Dir'iyyah Institute, Geneva (Atlanta, Georgia Institute of Technology, January 1978), pp. 26, 58–60.

Sometimes when social amenities are perceived as being localized in nature, that is, connected with special local customs or attributes, they are apt to be termed by outside analysts as "cultural constraints." Be that as it may, the principle is the same: the society in question has different social preferences than the energy analyst. Scattered reports of negative attitudes in China toward biogas and in Mexico toward solar cookers could reflect a local preference for "modernity" in rejecting the use of "village" solar energy sources to replace high-technology fuels. The attraction of some psychic amenities can appear obscure to those not familiar with local customs, but some have a more familiar basis. One taboo of a familiar kind is encountered in the use of animal (or human) wastes for generation of biogas. This problem can be found institutionalized in some caste systems in South Asia, but is not unknown elsewhere. The preferences of the Mbere tribe in Kenya for particular types of wood for different end-use purposes[30] may represent, on the contrary, fine distinctions in the actual burning attributes of different woods. Indeed, one may note an ancient and rational preference for hardwoods for fireplace use in industrial countries.

It cannot be too strongly emphasized that many of these social constraints are only really "problems" if we forget the total context of development—if it is supposed that the only goal of energy policy in a developing area is to secure least cost supplies of energy for appropriate end-uses. But in fact, such an outlook would not be compatible with the value preferences of most communities.[31] Some energy planners will view some of these local socioeconomic context problems with sympathy and others with less enthusiasm. Nevertheless, if the agreed-upon goal of a society is to carry out economic development so as to satisfy key social preferences, all questions of social and cultural constraints will have to be recognized as aspects of those preferences and therefore an essential component of the energy problem in each individual society.

Equity and Income Distribution

Equity or fairness questions are also of critical importance in any analysis of energy problems. Unfortunately, there are no objective

[30] David Brokensha and Bernard Riley, "Forest, Foraging, Fences and Fuel in a Marginal Area of Kenya," Paper presented at U.S. Agency for International Development Africa Bureau Firewood Workshop, Washington, D.C., June 12–14, 1978, p. 3ff.

[31] William Ramsay and Elizabeth Cecelski, "Energy, Scale, and Society," draft (Washington, D.C., Resources for the Future, 1980).

criteria to determine how fair is fair enough, or analytical methods to measure degrees of fairness even in a philosophical sense. Furthermore, there are few agreed upon policies that will successfully bring about more equitable income distribution in the general society. Thus, it is questionable whether energy policies are the right tools to accomplish improvements in income distribution. Even more difficult, how much equity should be purchased at the cost of a higher priced energy technology? Is it even possible to choose new energy strategies so that the poor become less poor? Or can one at least ensure that existing income problems are not made worse by new energy choices?

There are very few data in either industrial or developing areas on how energy supply options affect incomes and poverty levels. One of the few places where this question has been studied is in the field of rural electrification. In fact, one announced goal of rural electrification in many cases has been to raise the standard of living of the masses of people in rural areas.[32]

Analyzing the effects of rural electrification projects on various income groups is complicated by geography and population distribution. The more electrified regions tend to be those with larger populations, and larger population centers are more apt to be electrified than smaller. This biases the study population away from the lower income communities of rural areas. In fact, only 12 percent of the rural population in developing areas had access to electricity in 1971: 23 percent in Latin America, 15 percent in Asia, and 4 percent in Africa. Even these figures may be misleadingly optimistic: it has been estimated that only about 10 percent of the people in most "electrified villages" in India actually have electrical hookups.[33]

Despite this problem of geography, research in Latin America has shown a correlation of income with electricity connection, and a strong correlation between family income and levels of electricity use. But the other side of the coin is that consumption of some electricity does seem to begin, in two cases studied in El Salvador and the Philippines, at very low income levels. In one case in the Philippines, 67 percent of the "electrified" families had incomes below even official subsistence levels. To be sure, the electricity in this project was inexpensively priced and connections and appliances were subsidized.[34]

[32] For the following discussion, see Cecelski and coauthors, *Household Energy*, pp. 73–75.

[33] World Bank, *Issues in Rural Electrification*, Report No. 517 (Washington, D.C., 1975) p. 18; Amulya Reddy, "Energy Options for the Third World," *Bulletin of the Atomic Scientists* vol. 34, no. 5 (1978) p. 32.

[34] World Bank, *Costs and Benefits of Rural Electrification: A Case Study in El*

In any case, it must be kept in mind that such useful services and amenities as street lighting and refrigeration in stores and health clinics may accrue to low income classes as a community-wide benefit—although for street lighting an indirect income effect can arise if only richer neighborhoods are lighted.

The conclusion may well be that some rural electrification benefits accrue to the poor, but more of them tend to go to the relatively wealthy. Such a result might be expected for any economic good or service. But it becomes of special importance where supplies of rural electricity are sold at heavily subsidized prices: in such cases the urban poor may in effect be subsidizing the rural middle class.

Of perhaps even more concern is the introduction of new technologies that might divert energy or other resources from the poor to the rich. Any modern biomass-based technology might do this as an inadvertent by-product of other, more positive effects, such as raising rural incomes or employment. Alcohol industries could put pressure on land, diverting it from food crops or making direct encroachments on subsistence agricultural plots. Biogas generators are expensive—in the neighborhood of several hundred dollars—and require in Indian practice four or five cattle per family to supply them. Biogas digesters, therefore, tend to be an upper income technology, but their use could impinge on the supply of animal waste available to the lower income classes, depending on how the dung is obtained in the first place by the poor. This theoretical possibility, which could result in making dung into a cash commodity or in raising its price in existing markets through demand for biogas feedstocks—like the related problem of monetizing fuelwood—has been little documented.

In summary, one can see that—as is usual in economic analyses—the important questions of equity are difficult. Criteria are inexact and data on the connection between policy and incomes are usually poor. It is exceedingly difficult to identify the effects of the introduction of new technologies. One would expect that any technology requiring significant capital investment could act to increase the gap between rich and poor. On the other hand, a given technology might well provide direct benefits for some individuals plus community-wide benefits for all, so that absolute incomes tend to be raised for both rich

Salvador, P.U. Res. 5 (Washington, D.C., 1975) p. 74; National Electrification Administration, *Nationwide Survey on Socio-Economic Impact of Rural Electrification* (Manila, NEA, June 1978) pp. 18–21.

and poor. In the light of these uncertainties, a minimal planning position might be to ensure that new technologies should not depress the standard of living of the lower income classes.[35] While rural electrification appears to meet this test, it is at least conceivable that the promotion of new biomass technologies may require close attention if income distribution problems are not to be aggravated.

One common attack on the equity problem is to develop community energy systems that would involve a community pooling of costs and benefits. This approach, which really involves a restructuring of local society, has been briefly reviewed elsewhere.[36] As might be expected, such schemes present many theoretical and practical problems of organization and management, especially where community institutions are presently weak or nonexistent.

Conclusions

It is evident that the implementation of new energy technologies and the use of existing and new energy resources are not a matter of simple microeconomic planning. Obtaining capital for new energy schemes can be a big problem because often local financial markets are unable to handle increased needs for credit. Much can be usefully done to improve capital availability, however, by decentralization and other innovative banking techniques. Supplies of foreign capital may also be necessary in many cases to develop local energy resources, and arranging external credit can be a difficult national problem.

Essential infrastructure, such as roads that could make new energy technologies cost effective, may often be missing. The answer to this problem is to make overall transportation and water resource planning accommodate energy along with other development objectives.

Macroeconomic effects of energy systems such as impacts on national or local employment may be dominant over narrow cost considerations in planning practices. The importance of these concerns will be evaluated only through local political decisions.

Health and environmental effects must be monitored, even in countries with little development and many untapped environmental resources, if future development is not to produce unwelcome envi-

---

[35] This viewpoint is restricted in scope to energy decisions—alternative social investments might of course provide better amelioration of equity problems.

[36] Cecelski and coauthors, *Household Energy*, chapter 5.

ronmental surprises. However, energy planners in developing areas can make use of a growing literature in industrial countries on these externalities.

Proposed changes in energy use patterns will often be in conflict with many current local life-styles. Indeed, if all socioeconomic factors are considered, some new technologies may not reflect actual social preferences. They may then turn out to be costly in a total social sense, and some special technologies such as solar cookers may be inappropriate even when apparently cost effective on the basis of energy-unit costs.

Finally, the question of income distribution and equity, as always a difficult one, must be examined closely when energy changes are made. It may be impossible to trace all the implications of new technologies for criteria of fairness, much less to judge equity versus efficiency. Nevertheless, actual worsening of income gaps is a phenomenon that deserves attention: modernization of energy sources could widen income class gaps in some free-market environments. Community energy alternatives could in principle supply one answer to equity problems. But, as is often the case for collective development schemes, the required degree of community cooperation may not be feasible.

10

Energy Strategies and Policy Suggestions

In the opening chapter, we described today's world energy problems as a combination of short-term crisis and long-term transition. For oil-importing developing countries, the need in the short term is to cope with sharply higher oil import bills through some combination of adjustments in other imports and exports, oil conservation and sub-stitution, and international private or public financing of the increased trade deficits—at the same time minimizing the adverse impact on continuing development. For the longer term, the need is to accomplish an orderly transition to an altered régime of energy supply and use involving higher relative costs, different resource demands, and pos-sibly significant modifications in development strategies.

The consequences of mismanagement in either the short or long term may be catastrophic for economic growth prospects and for social and political stability. Nor can long-term problems be deferred until those of the short term are fully resolved. The lead times inherent in altering the basic character of the energy régime may result in costly penalties for failure to start as soon as possible. Therefore, the development of a comprehensive energy strategy has now become an imperative for developing-country governments—not as an isolated exercise, but as an integral part of their general economic management. Ideally, the framework for such a strategy should include a range of alternative scenarios for each country's energy future a generation hence, with a specification of the main policy options, the measures required to move from here to there, and the time periods in which firm decisions are needed. For some long lead-time issues, it may be necessary to take decisions on the basis of partial knowledge, but in many cases options can and should be kept open pending results of resource exploration, technological research, and field experimenta-

241

tion. Energy planning must therefore be a continuous process that does not seek to freeze the future.

The diversity of resource endowments and development conditions is so great that no single strategy could be appropriate for all countries. We outline in this chapter a series of broadly applicable suggestions for improving energy efficiencies, increasing energy supplies to replace imported oil, and fitting energy into broader development objectives. In some countries, action along many of these lines is already under way, and there are probably no cases for which all of them are relevant. But we believe that they can serve as a useful checklist for the developing-country analysts and policy makers who must carry the main burden of devising strategies to cope with the issues of energy crisis and transition at the national level. In addition, we set forth a number of suggestions for complementary policy measures intended for international institutions and industrial-country agencies concerned with energy and development. Beyond these kinds of measures lies the critical question of whether worldwide interest in a peaceful and effective energy transition could be made the basis for a genuinely global negotiated energy strategy.

Improving Energy Efficiency

Although developing countries consume much less energy than industrial countries on a per capita basis, there are important possibilities in these countries for using energy—in particular petroleum—more efficiently without adverse effects on economic growth or welfare. The suggestions are:

Know how much energy is being used and where. The first step in using energy more efficiently is to know how and where it is being used. This involves drawing up energy balance sheets so that major end-use sectors and substitution possibilities can be identified. The traditional fuel sectors must of course be included.

Improve energy "housekeeping" practices. "Housekeeping" practices such as regulating thermostats and temperatures in industrial processes are relatively simple and are an easy source of nearly costless energy savings once attention is drawn to waste. The small use of energy for space heating rules out some of these possibilities in developing countries but the others, including industrial uses and air conditioning in commercial buildings and hotels, are often substantial.

Plan for energy-saving technologies in new investments. Rising energy prices will make it economical to apply some energy-saving technologies in the short run, and many more in the long term. Because of its nearly complete dependence on liquid fuels and because it absorbs a large part of total oil consumption, the transportation sector is a prime candidate. More efficient vehicles, more effective public transportation, and a greater exploration of the wide variety of public and private transportation modes—short of highly mechanized, capital- and energy-intensive forms—should be undertaken.

The industrial and modern residential sectors of most developing countries are small but rapidly growing energy consumers, making them good targets for adoption of newer, more energy-saving industrial technologies and home appliances as they become available. Where the nature of energy consumption in developing and industrial countries is similar—as in parts of the industrial, residential, and transportation sectors—energy-conserving technologies can be borrowed. Many of these technologies are already known and have been widely used in countries with relatively high energy prices in the past; the challenge is to improve and refine these techniques, based on national circumstances and needs.

But a note of caution is in order. Energy-saving technologies developed in the industrial countries are likely to be highly capital intensive. Appropriate energy-*saving* technologies for the developing countries, no less than appropriate energy *supply* technologies, should reflect the relative social costs of capital and labor. Government action may be needed to stimulate research and development in this direction.

Increase the efficiency of traditional fuel use. Improving the efficiency of traditional cooking stoves, even by subsidy if necessary, could be one of the simplest and cheapest ways of providing more energy to the traditional sector without straining either fossil or traditional fuel supplies. However, the total costs, even at as little as $5 a stove, might still be high for an all-inclusive national program and would require careful planning and a commitment that goes beyond pilot projects.

Give correct energy price signals to consumers, with due attention to considerations of equity. Appropriate pricing involves carefully reexamining subsidies and taxes on energy consumption, many of which were developed under different conditions for purposes other than energy conservation and some of which may have perverse effects on energy supply and use. Most countries have taxes on certain energy supplies, in particular gasoline. Kerosine, on the other hand, is frequently sold at subsidized prices below its real economic opportunity cost. All energy-pricing structures, including electricity rates, should be re-examined in the light of rising energy costs.

Serious equity problems are raised by the prospect of higher energy prices, especially for the household and transportation needs of lower income groups. There is clearly a sharp conflict here between economic efficiency and equity. In principle, these needs should be addressed by methods of supplementing incomes outside the energy sector, but practical and effective methods may not be available and some compromises will be inevitable. It is highly desirable, however, for budgetary as well as efficiency reasons, that increased emphasis be given to energy pricing that reflects replacement costs of energy used.

Use taxes, subsidies, and regulations to achieve conservation—but with caution.   In addition to reforms of energy pricing, specific measures may be needed to secure conservation objectives. Typical examples are loans and subsidies to consumers and regulation of the energy efficiency of major energy-using appliances. Whatever supplementary measures are chosen to back up higher prices, care should be taken to minimize the additional burden on an already strained bureaucracy. Administrative simplicity should be emphasized, modifying existing procedures rather than instituting wholly new ones.

Increasing Energy Supplies

Fossil fuels and conventional hydroelectricity will continue to be the principal sources of commercial energy for at least the next two decades. The largest supplies of "new" domestic energy for developing countries are likely to come from conventional resources that were previously unexplored or unexploited—oil, gas, coal, and hydropower. Newer energy sources, such as nuclear power and geothermal energy, should be viewed with caution but could be significant for some countries. Biomass has important potential for expanded or more efficient production and use in both the traditional and modern sectors. Various "appropriate" technologies may have practical roles to play, but are still subject to unanswered technical and social questions.

Fossil and Uranium Fuel Resources

For those countries which possess the basic resources, the easiest road to easing national energy crises lies through finding and producing conventional hydrocarbons: oil and gas, coal, and shale. To accelerate the discovery and availability of these resources, new types of contractual arrangements can be negotiated between host country governments, private companies with the requisite technology and skills, and international financial institutions.

Encourage oil and gas exploration. Although there are conflicting views about the amounts of oil and gas left to be found in the world, there are substantial regions in the developing countries that have been little explored. Appreciable amounts of oil and gas will undoubtedly be found and will prove to be invaluable energy bonuses to countries finding them. A considerable amount of previously flared gas may also be converted to industrial use or electricity generation or liquefied for export.

Search for other fossil resources. A number of countries have coal or coal-like reserves that are relatively cheap to explore. Other fossil resources, especially oil shale, may require special exploration programs. It is almost certain that existing estimates of the total scope of these resources are low by a wide margin.

Examine carefully the nuclear power option. At present, an extensive commitment to nuclear power might be a risky energy option for developing nations. Lack of operating reliability and possible safety risks have become serious concerns of late for light water reactors. Moreover, these risks are of special concern in countries where technical cadres to manage an extremely complex and potentially dangerous technology may be scarce, and where limited electrical demand may mean that important economies of scale can be lost. World concern about proliferation, which could limit access to nuclear technologies and fuels to developing countries, is another complicating problem. Though intensive efforts are being devoted to improvements in safety and to arrangements for assured supplies to countries undertaking nonproliferation commitments, risks still remain. If a developing country decided to adopt the nuclear option now, it might benefit from a buyer's market, but it is a buyer's market for good reason.

Another less venturesome possibility for countries with uranium resources in economic concentrations is to develop them for export, whether or not these countries envisage the development of domestic nuclear power in the visible future. In those cases, uranium sales, like those of any other valuable export mineral, assist the balance of payments.

Renewable Energy Resources

The competitive position of conventional renewable resources—and some less conventional ones such as geothermal—has greatly improved as the result of higher oil prices. This leads us to the following suggestions:

Evaluate renewable energy resources. A general assessment of potential renewable energy supplies can help to identify some of the most promising resource options. Restraint should be exercised, however, to avoid the excessive human and budgetary costs of seeking complete data on all possible forms of renewable energy.

Re-examine national hydroelectric potential. Unimplemented hydroelectric schemes, deferred in the past because they could not compete with cheap oil, should now be reexamined. Small-scale "mini-hydro" power could also make a contribution, but its current vogue should not be allowed to overshadow the potential for larger scale conventional hydroelectric development, which in some regions is very great.

Move conservatively on geothermal power. A cautious attitude is advisable for most forms of geothermal power, pending the findings from costly experimentation in more affluent countries. An exception is the use of dry steam and high-temperature wet steam resources, but they have been discovered in only a few locations. The more widespread resources of lower temperature wet steam, liquid, and dry rock resources may be usable in the future, and research progress in prototype plants in industrial countries should be watched closely for developments.

Make vigorous efforts to apply biomass in the modern sector. Perhaps the most promising option is the use of biomass in generating electricity. In many developing-nation locations, even long-neglected natural forests could be systematically managed and harvested on a renewable, sustained yield basis, or more forest wastes gathered, to produce electricity at competitive costs under present conditions. By setting up tree plantations, significantly better yields per hectare can be obtained. Agricultural wastes, urban wastes, and the more exotic energy crops also often represent possible new concentrated sources of underutilized biomass.

Explore the alcohol fuel option.   The other key modern area is that of liquid fuels, particularly for transportation. One such fuel is methanol derived from forest resources, but this technology is still subject to engineering and economic uncertainties as well as unevaluated health risks. A less risky option, but also a more expensive one under foreseeable technological projections, is to grow special sugar and starch crops for conversion into ethanol. Ethanol has some advantages that may partly offset its expense: it can be used to replace gasoline and other petroleum products in motor vehicles. It is the equivalent of high octane gasoline, and can therefore be priced to some extent as a gasoline additive. Indirect benefits such as job creation and foreign exchange savings can also help to justify relatively high costs for alcohol, within the limits of realistic shadow pricing for the resources involved.

Increase biomass supplies for traditional uses.   It seems plausible that, with ever shifting energy economics, underutilized distant forest resources could be used more often for fuelwood or converted to charcoal to supply traditional open fires or stoves. Sustained yield management technologies are an essential element of such a strategy. In theory, market incentives should suffice to bring additional fuelwood to consumers, but institutional limitations may often require some government stimulus to bring about harvesting and transportation, as well as the development of new resources through tree plantations and community and private woodlots.

Be sure "appropriate" technologies are genuinely appropriate. Energy technologies should be appropriate in terms of true social costs: supplies of needed types of labor, capital, and other inputs, socioeconomic context, and sociopolitical goals. Where resources are limited, experimental technologies with little short-term promise from a technical or economic standpoint are better left to industrial countries to standardize and make reliable. For developing countries to build costly solar thermal electric demonstration plants, for example, may waste not only money and time but also limited scientific and engineering manpower, with disappointments for national expectations at a time when the need for serious energy alternatives is keenly felt.[1]

---

[1] Indeed, some of the more experimental renewable technologies could—ironically—turn out to be good examples of the dictum popularized by Amory Lovins, "Technology

Other promising experimental technologies—including solar options like photovoltaics—may need a broader approach than is implied in the widely promoted emphasis on decentralization and smallness of scale. One of the most efficient of the newer "appropriate" village technologies appears to be biogas, but it is probably most economical on a large scale and most practical as a commercial rather than a household or small community venture. More support should be given to the promotion of biogas and other renewable energy facilities as large community efforts or even as industrial enterprises.

Review village energy technologies in relation to national planning goals. "Village technologies"—those appropriate to village resources and capabilities—are often held to provide social value that goes beyond economics alone. While small-scale hydropower may be cheaper in some cases than power from a centralized grid, in other cases planners would need to count in other types of social benefits to counterbalance economic disadvantages. The use of "village technologies" to promote such ends as improved quality of life, a more equitable distribution of income and resources, and the generation of rural employment generally involves some economic costs. So far as possible, the economic price of broader objectives should be explicitly evaluated and deliberately assumed, including in the evaluation alternative (and perhaps less costly) means of achieving those social objectives.

Energy and Development

The previous suggestions focused on measures to improve energy efficiencies and to increase energy supplies in the developing countries. This section deals more broadly with the relationship between energy and economic development, in particular with how energy policies should fit into overall development strategies. In general, energy policy

---

is the answer, but what was the question?" To be sure, for those technologies which show real promise, local pilot operations will be needed in due course. Indeed, in the case of energy supplies from biomass, where local soil and climatic conditions may be decisive factors in determining yields and costs, field experimentation in the developing countries themselves is indispensable.

should be subordinated to broader development goals, but it should be recognized that the changing economics and availability of energy make it a potentially critical consideration in some national development policies. These suggestions are:

Integrate energy sector planning into broader development strategies. The jump in energy prices and uncertainties in the availability of oil have produced a universal recognition of the need for comprehensive energy planning as part of every country's development efforts. Nonetheless, energy supply and demand ought not to be the dominant feature of development planning, or something to be planned in isolation from broader national objectives. The only rational policy is to put energy planning in its properly subordinate position, while recognizing that the changing economics of energy will entail new constraints affecting the overall strategies.

In some planning areas—urbanization, transportation, industrial development, integrated rural development, regional development, and technology policies—energy could be an especially important factor. In each of these areas plans may need to be revised in the light of higher energy costs or shifts in desirable forms of energy supply. At the very minimum, the energy implications of existing plans should be reviewed.

Do not seek energy self-sufficiency regardless of cost. Abrupt price rises and supply uncertainties have provided more than adequate reason for nervousness about energy on the part of economic planners during the last decade. Many countries feel that the burden of oil imports has become intolerable and are seeking ways to achieve a significantly higher degree of energy self-sufficiency by subsidizing and otherwise encouraging domestic production. But energy self-sufficiency comes at a price, and in many cases the price will be prohibitively high. Alternative means to provide security against supply disruption, such as diversification of sources and stockpiles, should be examined. On the economic side, a less costly alternative may be the expansion of exports to pay for at least some level of continuing energy imports. All energy self-sufficiency subsidy proposals should be scrutinized with care to make sure that the costs do not outweigh the benefits. This assessment should of course include indirect benefits, including such external effects as increased employment of underutilized labor, a possible premium for foreign exchange savings, and wider latitude in the making of foreign policy.

Supportive Policies by International Institutions and Industrial Countries

The industrial countries play critically important roles with respect to the developing nations, in energy as in other fields. Mechanisms for assisting in the solution of developing-country energy problems already exist in the bilateral aid programs of several industrial countries and in the activities of the major foundations. In addition, the multilateral aid-giving organizations such as the World Bank, the regional development banks, and the U.N. Development Programme are planning to expand their work in this area, perhaps with special contributions from OPEC member countries. A U.N. conference on new and renewable sources of energy, scheduled for 1981, will provide a vehicle for stimulating additional efforts and will draw attention to the need for more initiatives from the developing countries themselves, who are in the best position to assess their own energy needs and to specify the types of assistance required.

Complementary actions by international institutions and industrial countries could reinforce the policies suggested for developing countries in a number of respects.

Help the developing areas find more fossil and other mineral resources.   The institutional and informational barriers to finding new resources are probably not fully appreciated. For petroleum, where extensive exploration has taken place in the past, a recent World Bank program has recognized the need for special aid to developing nations.[2] For coal, oil shale, and perhaps uranium, official assistance in organizing technical and planning help may be needed, especially because of the absence of experienced multinational enterprises accustomed to working in a great variety of physical and political environments.

Help developing countries reassess the hydroelectric potential.   In many nations, technical help could be well used in reassessing hydroelectric possibilities in the light of changing relative costs. The potential for small-scale hydroelectric power has only been touched and could benefit from large amounts of engineering and economic analysis help from abroad.

Help with developing-country forestry schemes.   International technical assistance in forestry has received relatively little attention, even though the application of modern sustained yield management practices could greatly increase the contribution of forests to devel-

[2] World Bank, *Energy in the Developing Countries* (Washington, D.C., 1980).

opment needs as well as help to discourage deforestation. Helping to optimize the vast biomass potential in developing areas can be good business for all nations, for use both in energy supply and in wood products.

Ensure the transfer to developing countries of alcohol fuel research results. Much research will be carried out on gasohol and related problems in the United States and other industrial countries. The results could be of great value to developing areas interested in ethanol (or possibly methanol) as a substitute liquid fuel. In this area, arrangements to share the experience of one newly industrializing country, Brazil, may be especially indicated.

Coordinate coal synfuels development with biomass feedstocks. Much of the synthetic fuel development in the United States, Britain, and Germany will be carried out in terms of coal feedstocks. However, because the technologies for coal and biomass may be very similar, a small additional effort more specifically directed at biomass feedstock possibilities would have the dual advantage of (1) preparing the United States and other industrial countries for the later development of domestic biomass supplies and (2) providing help now for biomass fluids programs in developing areas.

Do not look upon experimental energy solutions as practical energy assistance. Experimentation with new energy sources, such as wave power, ocean thermal energy conversion, or solar power towers, are entirely appropriate activities for an industrial country. They may represent significant energy sources of the future. But the international institutions and industrial countries should avoid the premature introduction of untested technologies into developing areas. Besides wasting resources, there is a danger that the demonstration may contribute to local misapprehensions of realistic energy options.

Clarify the nuclear issues. For the developing countries with a serious interest in nuclear energy, difficulties are posed not only by technical and economic problems, such as costs, safety, and waste management, but also by uncertainties concerning the policies of the nuclear supplier countries, especially in relation to weapons proliferation concerns. The first set of problems is shared with industrial countries and is an unavoidable consequence of the still fluid state of nuclear technology. The second set could be greatly alleviated by clarification of policies in the wake of the International Nuclear Fuel Cycle Evaluation of 1977–1980. In particular, stable and consistent rules are needed on the conditions governing supplies of nuclear fuel and enrichment services, spent fuel storage, and requirements for international safeguards. Developing countries should be represented in the negotiation of such rules.

Analyze existing aid programs for energy implications. Donor countries should analyze their existing aid programs of all types for any implicit bias toward energy-intensive technologies. While such technologies may still be the optimal choice, as frequently emphasized in this study, an identification of energy implications is essential. In addition, donor countries should expand their programs of energy assistance.

Worldwide Energy Cooperation

In analyzing the position of the developing countries in the global energy framework at the beginning of this study, we pointed to the worldwide interconnections among all forms of energy production and use and the basic mutuality of interest among industrial and developing countries in a peaceful transition from the era of cheap oil to a more durable energy economy sufficient to the needs of all. A smooth transition is obviously in the interest of both industrial and developing oil importers. It should also be of major concern to exporters, whose sources of supply and markets require some degree of world stability and who must consider their own ultimate transition to economic diversification and substitute energy sources. The alternative of competitive scrambles, bilateral "special deals," and threats or use of military force to secure supplies is extremely hazardous, even to those who suppose that they might secure a temporary gain.

This shared concern about energy is more tangible and immediate than the general interest of richer countries in the reduction of poverty and economic progress in the developing countries. In the case of energy, all forms and sources of supply, demand, and conservation ultimately affect the balance of the international trade in oil, and oil has a unique impact on the system of international finance, the pressures of inflation, and the resources available for investment and development. A mismanaged energy transition is a potential bottleneck—even a source of potential strangulation—for continued economic growth in all parts of the world. But many aspects of oil production and use are not effectively governed by competitive market forces, and the speed and shape of the transition to a new energy régime will be enormously influenced by government policies.

In a world where political power is concentrated in sovereign nation-states, most of the responsibility for public policies necessarily rests on national authorities. But a cardinal issue of energy policies is how far they are to be independent and unilateral and how far harmonized through consultations among groups of nations or coordinated on a worldwide basis through informal understandings or formally negotiated agreements. In practice, the current situation is a mixture of national and international arrangements. In the case of oil, there is significant harmonization within OPEC on price policies. A very limited degree of coordination exists among the principal industrial-country importers, through the International Energy Agency, on subjects such as conservation, import levels, development of supply alternatives, and contingency planning for emergency shortages. On nuclear proliferation, there is organized cooperation in the International Atomic Energy Agency and the London Nuclear Suppliers' Group, and there are likely to be new institutional arrangements in the wake of the International Nuclear Fuel Cycle Evaluation.

The developing-country energy importers have not yet organized themselves for the pursuit of a joint energy strategy. They are excluded from OPEC by definition and have had only occasional liaison with the International Energy Agency. At the regional level, there are useful but very limited arrangements in Southeast Asia and in Latin America for exchange of information and technical experience. Yet the underlying convergence of interest and the growing importance of this group of countries on the world energy scene point to the need for their active collaboration, both with one another and with other groups.

A smooth energy transition ensuring adequate supplies to all countries is a classic case of a worldwide "public good." Like the frequently cited example of national defense, it cannot be broken up into private shares and traded in markets. The international oil price is sensitive to actions anywhere that conserve energy, find new sources of conventional fuels, or develop innovative renewable resources at competitive costs. But if some countries help to relieve pressures on that market through massive and costly programs of conservation and substitution, others will reap many of the benefits. In these circumstances, there is a temptation for each country to hold back in hopes of getting a "free ride," even though all would gain by moving forward in step. A common fear in developing countries is that they are being urged to conserve or to rely exclusively on renewable sources so that the world's dwindling hydrocarbon supplies can be kept as a preserve of the already industrialized countries. Furthermore, developing and industrial countries alike are confused and apprehensive about the upward spiral of oil prices, and about the issues of coping with basic problems or uncertainties in both short-run and future supply planning.

These are the types of considerations, coupled with the economic and security dangers of a scramble for energy resources, that are leading to widespread suggestions for development of a global energy strategy. One recent example is the Independent Commission on International Development Issues, chaired by former West German Chancellor Willy Brandt, which has called for worldwide collaboration on several fronts to promote an orderly energy transition.[3] Its main proposals comprise commitments by oil exporters to maintain assured production levels, commitments by all major energy consumers to agreed conservation targets and standards, and agreements by both exporters and importers on oil prices including indexation against inflation and guarantees for investments of oil surplus countries— together with financial assistance for developing-country oil imports in the short term and major investments in longer term fossil fuel development and energy research generally.

A critical appraisal of these and other proposals for worldwide energy cooperation would go far beyond the bounds of this study. Tensions within the exporting and importing groupings, as well as between them, may make the idea of global agreement a utopian dream. Yet the stakes are extremely high, and it would be unwise to leave unexplored any possibility for fruitful international coordination of energy strategy. To achieve their full potential, any arrangements that may be worked out must provide for the full participation of the developing countries.

[3] *North–South: A Program for Survival,* Report of the Independent Commission on International Development Issues (Cambridge, Mass., MIT Press, 1980) pp. 160–171, 278–280.

Conversion of Energy Units

For energy units that are physical identities, conversion factors carried to three significant figures are used. The most relevant to this book are:

1 kWh (kilowatt-hour) = 3.6 million joules (or 0.0036 gigajoules [GJ]) = 3,412 Btu (British thermal units)

1 GJ = 948,000 Btu (or 0.948 million Btu)

1 million Btu = 1.055 GJ

Crude oil occurs in a wide range of specific gravities (densities), with consequent differences in the weight of the standard barrel (42 U.S. gallons). The range is from 6.98 to 7.73 barrels per metric ton. The midpoint is about 7.35; the U.S. Department of Energy uses 7.33; but 7.3 has the practical advantage of making 50 million metric tons per year equal to 1 million barrels per day. Using that factor yields the following equivalents:

1 metric ton (tonne) of oil = 7.3 barrels of oil

50 metric tons per year = 1 barrel per day

1 million metric tons per year = 20,000 barrels per day

The heat content of both *coal* and *oil* vary greatly, and the authorities differ in their use of conversion factors. The U.N. Statistical Department is now publishing its global data in metric tons both of coal equivalent (as in the past) and of oil equivalent. It has adopted standard conversion factors of one metric ton of coal = 0.680272 metric tons of oil = 29.3076 GJ. Those coefficients imply a heat content of oil of 5.6 million Btu per barrel, which is close to the average consumption figure for petroleum products used by the Department of Energy although somewhat lower than the average for crude oil production (5.8 million Btu per barrel). The factors used in this book, based on U.N. practice, are as follows:

1 metric ton of coal = 0.680 metric tons of oil = 29.3 GJ = 27.8 million Btu

1 metric ton of oil = 1.47 metric tons of coal = 43.1 GJ = 40.8 million Btu

1 barrel of oil = 5.90 GJ = 5.60 million Btu

Natural gas also varies greatly in heat content. The weighted average used by the United Nations is 9,320 kilocalories per cubic meter, which converts to 37,775 Btu per cubic meter or 1,070 Btu per cubic foot. This is slightly higher than the Department of Energy figure of 1,021, but again has the advantage of uniform practice for global data. On the U.N. basis, the conversion equivalents are as follows:

1,000 cubic meters of natural gas = 1.359 metric tons of coal = 0.926 metric tons of oil = 39.85 GJ = 37.8 million Btu.

Hydroelectricity and *nuclear energy* both confront the well-known problem of whether the fossil fuel equivalent should be measured in terms of the electricity output or the equivalent input of fossil fuels that would be required in a thermal power plant to generate the same amount of electricity. The choice makes a large difference, since about two-thirds of the fossil fuel heat input in a power plant is lost in the condensed steam. The heat equivalent of the electrical output is a physical identity, as already noted (1 kWh = 0.0036 GJ = 3,412 Btu). The input equivalent varies with the fuel used and the efficiency of the power plant, which is measured by the "heat rate." The Department of Energy currently estimates the appropriate heat rate at 10,435 Btu per kilowatt hour. The U.N. data use the electrical output rate for all conversions, but most RFF studies have preferred the input rate, which avoids misleading comparisons of energy intensity between countries with large and small (or zero) shares of hydroelectric and nuclear power in their total supplies. This book uses the input rate unless there are specific reasons for using the output rate in a particular context, in which case the use of that rate is explicitly noted. The conversion factors are as follows:

Input rate:
1 million kWh = 10,435 million Btu = 11,009 GJ = 376 metric tons of coal = 255 metric tons of oil = 1,865 barrels of oil

Output rate:
1 million kWh = 3,412 million Btu = 3,600 GJ = 123 metric tons of coal = 83.5 metric tons of oil = 610 barrels of oil

Index

Joy Dunkerley has devoted the past ten years to research and writing on energy conservation. Now a senior fellow at Resources for the Future, she was a staff economist on the Ford Foundation Energy Policy Project which produced the first comprehensive set of energy consumption scenarios for the United States. At RFF for the last half-dozen years, she has completed three studies on the energy conservation lessons to be drawn from comparing U.S. energy consumption patterns with those of other industrial countries. She currently is engaged in research on energy conservation possibilities in developing countries.

William Ramsay, after earlier research in high-energy nuclear physics and astrophysics, has studied the systems analysis of village economies, urban land use, and waste-management systems. From 1972 to 1975, at the Atomic Energy Commission, he worked on environmental aspects of energy production, and later, as technical adviser to one of the commissioners of the Nuclear Regulatory Commission, he dealt with nuclear proliferation, safety, and risk analysis. He joined Resources for the Future in 1976 and now is a senior fellow there. His research at RFF has focused on health and environmental costs of conventional and renewable energy sources and, more recently, on energy economic problems in developing areas.

Lincoln Gordon, a political economist, became a senior fellow at Resources for the Future in 1975. His prior career was divided between academic appointments and government service. On the academic side, he was a member of the Harvard University faculty from 1936 to 1961, serving as William Ziegler Professor of International Economic Relations from 1955 to 1961. From 1967 to 1971, he was president of The Johns Hopkins University, and from 1972 to 1975 he was a fellow of the Woodrow Wilson International Center for Scholars at the Smithsonian Institution. On the government side, appointments included director of the U.S. Operations Mission and minister for economic affairs in the American embassy in London from 1952 to 1955; U.S. Ambassador to Brazil from 1961 to 1966; and Assistant Secretary of State for Inter-American Affairs from 1966 to 1967. His work at RFF has examined global economic and political trends, nuclear proliferation issues, and energy and development.

Elizabeth Cecelski was associated with Resources for the Future between 1977 and 1980. Her work there included research on energy and development, decentralization and appropriate technologies, nuclear proliferation, and international trade and development. She currently is senior energy adviser at Volunteers in Technical Assistance (VITA).